# Safety Management for Software-based Equipment

FOCUS SERIES IN COMPUTER ENGINEERING AND IT

*Series Editor Jean-Charles Pomerol*

# Safety Management for Software-based Equipment

Jean-Louis Boulanger

iSTE

WILEY

First published 2013 in Great Britain and the United States by ISTE Ltd and John Wiley & Sons, Inc.

ISTE Ltd
27-37 St George's Road
London SW19 4EU
UK

www.iste.co.uk

John Wiley & Sons, Inc.
111 River Street
Hoboken, NJ 07030
USA

www.wiley.com

Library of Congress Control Number: 2012955536

British Library Cataloguing-in-Publication Data
A CIP record for this book is available from the British Library
ISSN: 2051-2481 (Print)
ISSN: 2051-249X (Online)
ISBN: 978-1-84821-452-1

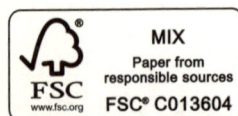

Printed and bound in Great Britain by CPI Group (UK) Ltd., Croydon, Surrey CR0 4YY

# Contents

# Introduction

Systems based on programmable electronics are being used increasingly widely and can make the tackling of safety even more challenging. Indeed, this type of system combines the strengths and weaknesses of computer and software programs. Electronics is characterized by faults, referred to as random, that can appear at any time but which can be predicted with probabilities. These are systematic faults (design, errors, misunderstandings, software faults, etc.). So software is not subject to ageing but to the notion of "bugs" (software fault). It can be argued that all software is subject to bugs and only software having undergone specific development processes may tend towards zero-fault.

The aim of this work is to describe the general principles behind the creation of a reliable programmable computer-based system. We shall outline the basic concepts for operating safety and the basic definitions (Chapter 1) as well as their implementation (Chapter 4). This book can apply to various normative situations (see Chapter 3), even though the example used is set in the railway field.

Tackling safety in a programmable computer-based system depends on mastering electronics (Chapter 5) and software (Chapter 6). In this type of system, it is to be noted that certification can be requested (Chapter 7).

To conclude this introduction, I would like to thank all of the manufacturers that have placed their confidence in me for more than 15 years.

# Safety Management

This chapter introduces the concept of system dependability (reliability, availability, safety and maintenance) and the associated definitions (fault, error, failure). One important attribute of dependability is safety. Safety management is in general related to people and everyday life. It is a difficult and high cost activity.

## 1.1. Introduction

The aim of this book is to describe the general principles behind the designing of a dependable software-based package. This first chapter will concentrate on the basic concepts of system dependability as well as some basic definitions.

## 1.2. Dependability

### 1.2.1. *Introduction*

First of all, let us define dependability.

DEFINITION 1.1 (DEPENDABILITY).– *Dependability can be defined as the quality of the service provided by a system so that the users of that particular service may place **a justified trust** in the system providing it.*

In this book, definition 1.1 will be used. However, it is to be noted that there are other more technical approaches to dependability. For reference, for the IEC/CEI

1069 ([IEC 91]), dependability measures whether the system in question performs exclusively and correctly the task(s) assigned to it.

For more information on dependability and its implementation, we refer readers to [VIL 88] and [LIS 95].

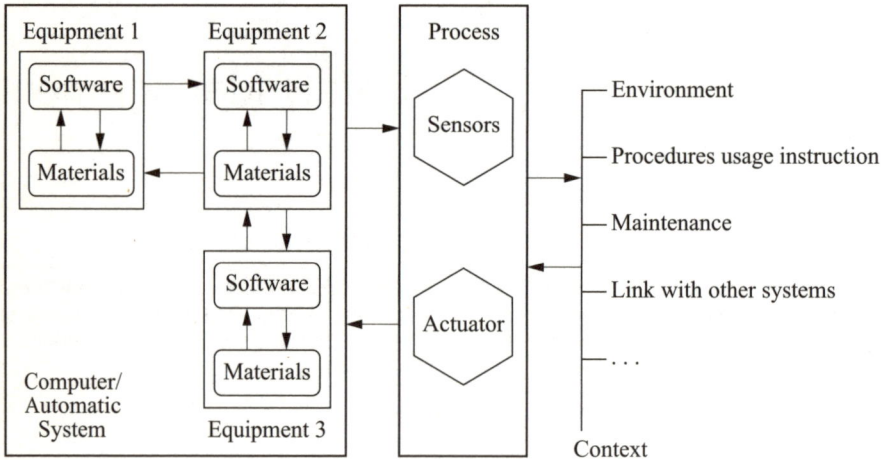

**Figure 1.1.** *System and interactions*

Figure 1.1 shows that a system is a structured set (of computer systems, processes and usage context) which has to form an organized entity. Further on, we shall look at software implementations found in the computer/automatic part of the system.

Dependability is characterized by a number of attributes: reliability, availability, maintainability and safety, as seen in RAMS[1].

New attributes are starting to play a more important role, such as security, and we now refer to RAMSS[2].

---

1 Reliability, Availability, Maintainability and Safety.
2 Reliability, Availability, Maintainability, Safety and Security.

### 1.2.2. Obstacles to dependability

As indicated in [LAP 92], dependability in a complex system may be impacted through three different types of event (see Figure 1.2): failures, faults and errors.

The elements of the system are subject to failures, which can lead the system to situations of potential accidents.

System 1                                                      System 2

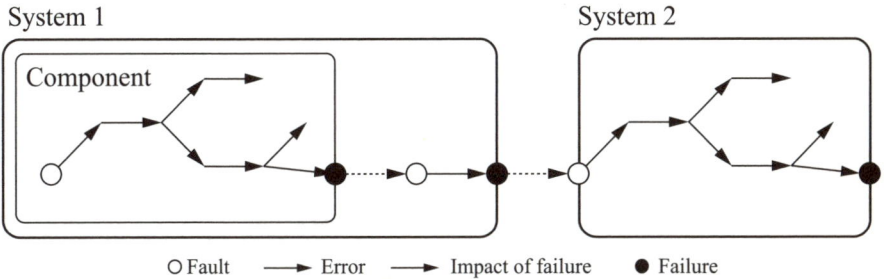

O Fault    ⟶ Error    ⟶ Impact of failure    ● Failure

**Figure 1.2.** *Impact from one chain to another*

DEFINITION 1.2 (FAILURE).– *Failure (sometimes referred to as breakdown) is a disruption in a functioning entity's ability to perform a required function. As the performance of a required function necessarily excludes certain behaviors, and as some functions may be specified in terms of behaviors to be avoided, the occurrence of a behavior to be avoided is a failure.*

From definition 1.2 follows the necessity to define the notions of normal (safe) behavior and of abnormal (unsafe) behavior with a clear distinction between the two.

Figure 1.3 shows a representation of the possible states of a system (correct vs. incorrect) as well as all the possible transitions among those states. The states of the system can be classified into three families:

– correct states: there are no dangerous situations;

– safe incorrect states: a failure has been detected but the system is in a safe state;

– incorrect states: the situation is dangerous and out of control; potential accidents.

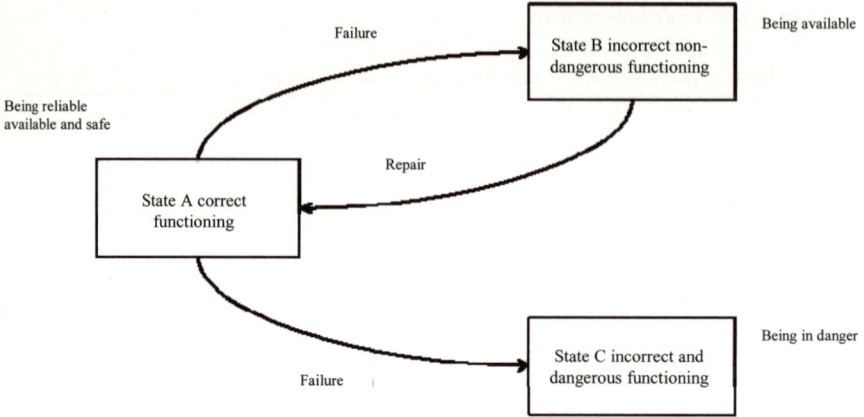

**Figure 1.3.** *Evolution of a system's state*

When a system reaches a state of safe emergency shutdown, there may be a complete or partial disruption of the service. This status may allow a return to the correct state after repair.

The failures can be random or systematic. Random failures are unpredictable and are the result of a number of degradations involving the material aspects of the system. Generally, random failures can be quantified due to their nature (wear-out, ageing, etc.).

Systematic failures are deterministically linked to a cause. The cause of failure can only be eliminated by a resumption of the implementation process (design, fabrication, documentation) or a resumption of the procedures. Given its nature, systematic failures are not quantifiable.

The failure is an observable external manifestation of an error (the standard CEI/IEC 61508 [IEC 08] refers to it as an *anomaly*).

DEFINITION 1.3 (ERROR).– *Error is an internal consequence of an anomaly in the implementation of the product (a variable or a state of the flawed program).*

In spite of all the precautions taken in the design of a component, this may be subject to flaws in its conception, verification, usage and maintenance in operational conditions.

DEFINITION 1.4 (ANOMALY).– *An anomaly is a non-conformity introduced in the product (e.g. an error in a code).*

From the notion of an anomaly (see definition 1.4), it is possible to introduce the notion of a fault. The fault is the cause of the error (e.g. short-circuiting, electromagnetic perturbation, or fault in the design). The fault (see definition 1.5), which is the most widely acknowledged term, is the introduction of a flaw in the component.

DEFINITION 1.5 (FAULT).– *A fault is an anomaly-generating process that can be due to human or non-human error.*

.... ⟶ Failure ⟶ Fault ⟶ Error ⟶ Failure ⟶ ....

**Figure 1.4.** *Fundamental chain*

To summarize, let us recall that trust in the dependability of a system may be compromised by the occurrence of obstacles to dependability, i.e. faults, errors and failures.

Figure 1.4 shows the fundamental chain that links these obstacles together. The occurrence of a failure may entail a fault, which in turn may bring one or more error(s). This (these) new error(s) may consequently produce a new failure.

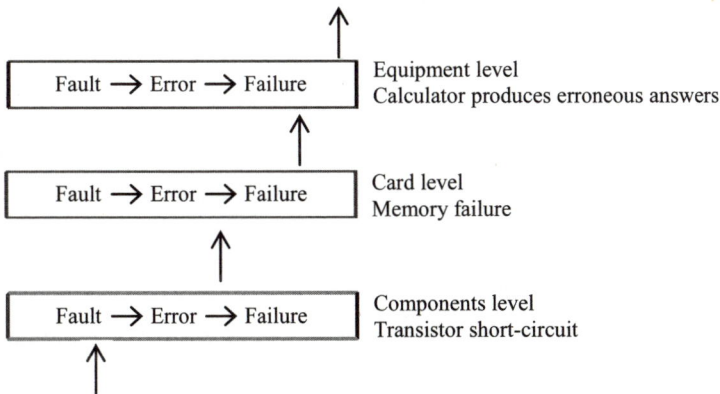

**Figure 1.5.** *Propagation in a system*

The relationship between the obstacles (faults, errors, failures) must be seen throughout the system as an entity, as shown by the case study in Figure 1.5.

The vocabulary surrounding dependability has been precisely defined. We shall henceforth only present the concepts useful to our argument; [LAP 92] contains all the necessary definitions to fully grasp this notion.

### 1.2.3. *Obstacles to dependability: case study*

Figure 1.6 illustrates an example of how failures can occur. As previously mentioned, a failure is detected through the behavior of the system as it diverges from what has been specified. This failure occurs at the limits of the system because of a number of errors, which are internal to the system, and has consequences on the working out of the results.

In our case study, the source of errors is a fault in the embedded executable. These faults may be introduced either by the programmer (a bug) or the tools (in generating the executable, downloading tools, etc.) or they can occur because of failures of the material (memory failure, short-circuiting of a component, external perturbation (e.g. EMC[3]), etc.).

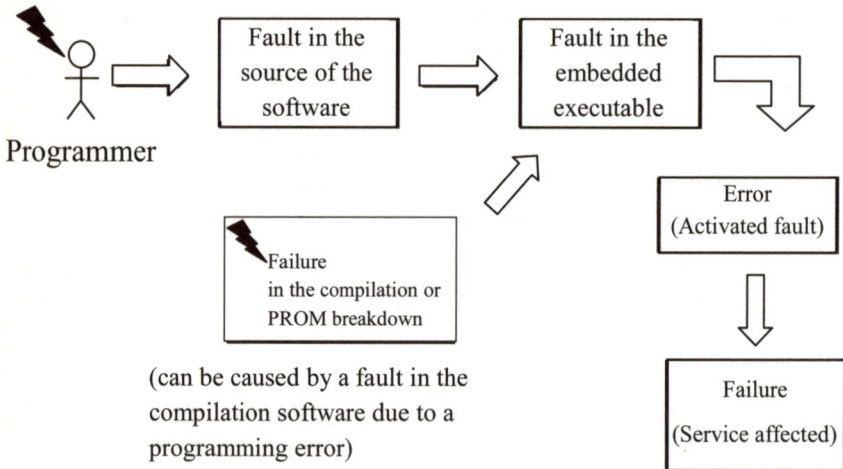

**Figure 1.6.** *Example of failure*

It is to be noted that faults may be introduced in the design (fault in the software, under-sizing of the system, etc.), in the production (when generating the executable,

---

3 EMC stands for Electromagnetic Compatibility, EMC is the branch of electrical engineering that studies unintentional generation, propagation and reception of electromagnetic energy.

in the manufacturing of the material, etc.), when installing, using and/or maintaining the software. The diagram in Figure 1.7 may then be declined for those various situations. Figure 1.7 shows an example of the impact of a human error.

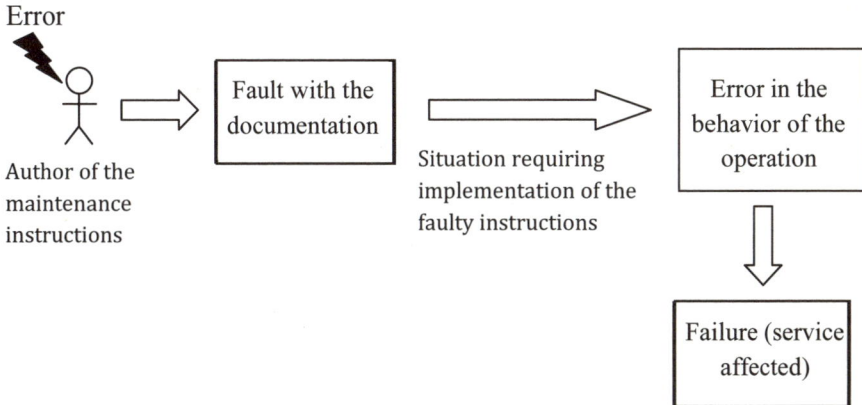

**Figure 1.7.** *Impact of human error*

At this point, it is interesting to note that there are two families of failures, i.e. systematic and random failures. Random failures result from the production process, age, wear-out, degradations, external phenomena, etc. Systematic failures can be reproduced as they originate in the design. Let us note that random failures may also result from a fault in the design such as underestimating the effect of temperature on the processor. As we shall see later, there are various techniques (diversity, redundancy, etc.) that can be used to detect and/or bring random failures under control. Systematic failures may pose a challenge as the issue of quality is involved and verification and validation are required.

### 1.2.4. *Safety demonstration*

The previous section clarified some of basic concepts (fault, error and failure). The systematic search for failures and their effects on the system is performed through activities such as preliminary hazard analysis (PHA), failure mode and effects analysis (FMEA), fault tree analysis (FTA), etc.

These types of analysis are now standard for dependability management and demonstration (see e.g. [VIL 88]) and imposed by the standards. All these analyses are used to construct a safety demonstration which is then formalized through a safety record: the safety case. The generic standard CEI/IEC 61508 [IEC 08], which is applicable to electronics and programmable electronics systems, covers this point and proposes a general approach.

### 1.2.5. *Summary*

When designing a software package, one should bear in mind three possible types of failure:

– random failures of the material components;

– systematic failures in the design, either material or software-based;

– specification "errors" in the system; these may have serious consequences on the operation and maintenance of the system.

### 1.3. Conclusion

In this chapter we have presented the basic notions related to dependability through definitions and examples. In the following chapters of this book, we shall show how to decline and take into account these notions and principles in order to make the system dependable in spite of the presence of a fault.

It is to be noted that with systems that impact on safety, standards are to be applied that encompass aspects of bringing failure safety and management under control.

### 1.4. Bibliography

[IEC 08] IEC, IEC 61508: Sécurité fonctionnelle des systèmes électriques électroniques programmables relatifs à la sécurité, international standard, 2008.

[IEC 91] IEC, 1991

[LAP 92] LAPRIE J.C., AVIZIENIS A., KOPETZ H. (eds), "Dependability: basic concepts and terminology", *Dependable Computing and Fault-Tolerant System*, vol. 5, Springer, New York, NY, 1992.

[LIS 95] LIS, Laboratoire d'Ingénierie de la Sûreté de Fonctionnement, *Guide de la sûreté de fonctionnement*, Cépaduès, 1995.

[VIL 88] VILLEMEUR A., *Sûreté de fonctionnement des systèmes industriels*, Eyrolles, Paris, 1988.

# 2

# From System to Software

Safety management is a continuous activity from the system to the software. From hazards at the system level, we can deduce the safety requirements and objectives (tolerable hazard rate and design assurance level). These safety requirements and objectives can be allocated on the subsystem up to the hardware and the software. This chapter presents this approach and the associated methodology.

## 2.1. Introduction

Numerous control systems (in the railway, aeronautical, nuclear and automotive sectors) as well as process control systems (production, etc.) are being been made to be automatic and computer-based. Moreover, logic and analog systems, little integrated by systems with high integration potential, are being replaced by computer-based systems. This has brought about a considerable broadening of the field of dependability including the characteristics and specificities of computer systems.

Dependability particularly affects implementations for which correct functioning is continuous and imperative, either because of the risk to human life (transport, nuclear energy, etc.), due to the high investment that would be lost following a failure in calculation (space, manufacturing process, process of chemicals, etc.) or even due to the cost and the disturbance incurred in the case of failure (banks, transport reliability, etc.). Over the past few years, there has been a realization of the impact of failures on the environment (accidental chemical spills, impact on ecosystems, recycling, etc.)

From the first stages of such systems, validation problems are the designer's main concerns. The mechanisms allowing a response to failures will have to be proven to be of good quality, this design will have to be tested (simulations, tests, evidence, etc.) and significant variables will have to be predicted to measure the performance of dependability.

The challenge is to correctly identify the various stakeholders (users, operators, managers, maintenance staff, authorities, etc.), the various elements of the system, the interactions between these elements, the interactions with the users and the elements having an impact on the dependability of the system. Finally, the electronic and/or programmable elements will have to be recognized.

This second chapter aims to put software in the context of its use as a system and to evoke the relationships that have to be considered when designing software.

## 2.2. Systems of command and control

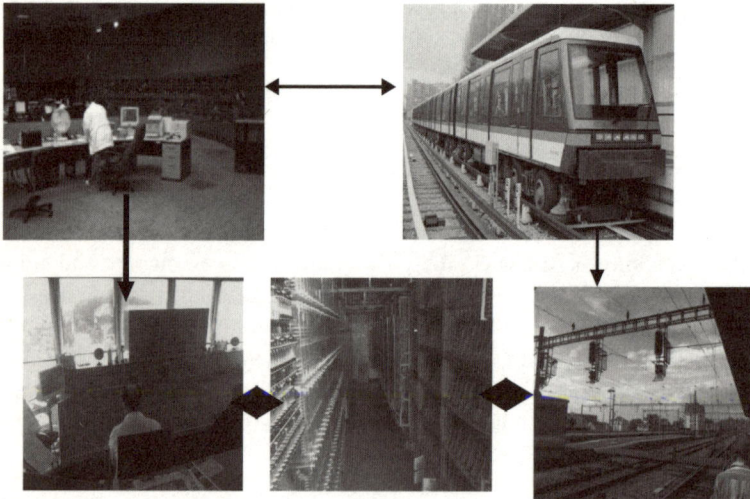

**Figure 2.1.** *The system in its environment[1]*

Figure 2.1 presents an example in the railway sector. The operation control center (OCC – top-left picture[2]) allows control over the whole line and the sending of commands to the trains as well as to the signaling management system (the bottom-left picture shows a manual operating control center).

1 Pictures by Jean-Louis Boulanger.
2 The picture shows an old-generation OCC; new OCCs are based in PCs and have developed from a physical technology (TCO – optical control view ) to display by a video projector.

Figure 2.1 introduces us to the complexity associated with the system and reminds us that a complex system is not based on one but on a multitude of software packages. Each package is associated with safety objectives that may diverge.

The central control room[3] sends commands to the field through a relaying system (the bottom central picture shows a manual operation's control center). The field packages adopt the required behavior in response to the commands (the bottom right picture shows the operating signals).

For instance, the software packages used for supervision do not impact as much on people's safety as those related to automatic train driving. This is why in the context of systems requiring an authorization for the service (aeronautics, railways, nuclear, programmed-computer systems, etc.), a safety assurance level is associated with each of the parts (equipment, subsystem, etc.).

This safety assurance level is measured on a scale going from "non-critical" to "highly critical". Notions of safety assurance levels and scales will be explained in Chapters 3 and 4.

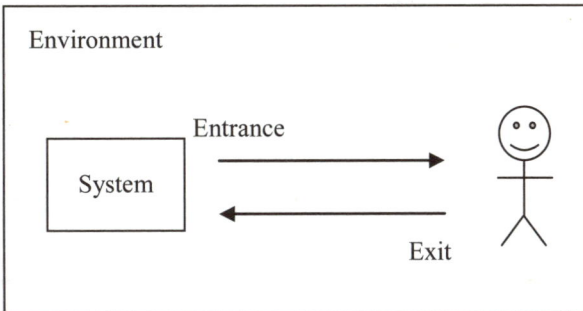

**Figure 2.2.** *The system in its environment*

Figure 2.2 shows that the system to be implemented interacts with its environment, which in turn reacts to the commands. It is therefore necessary to acquire an overall view of the process to be controlled, and to access to command tools allowing the transmission of the commands to the environment. The latter may include physical elements but generally, the interactions involve humans (operators, users, maintenance staff, etc.).

---

3 Figure 2.1 shows a center based on a relay in the middle picture. These have become computerized and are referred to as PMI ([BOU 11 – Chapter 8]), PIPC and PAING ([BOU 10 – Chapters 4 and 5]).

When the needs are analyzed, it is essential to correctly identify all the actors (operators, maintenance agent, client, etc.) and to identify all the tools that interact with the system. The analytic phase of the needs is crucial and remains the source of numerous omissions and misunderstandings.

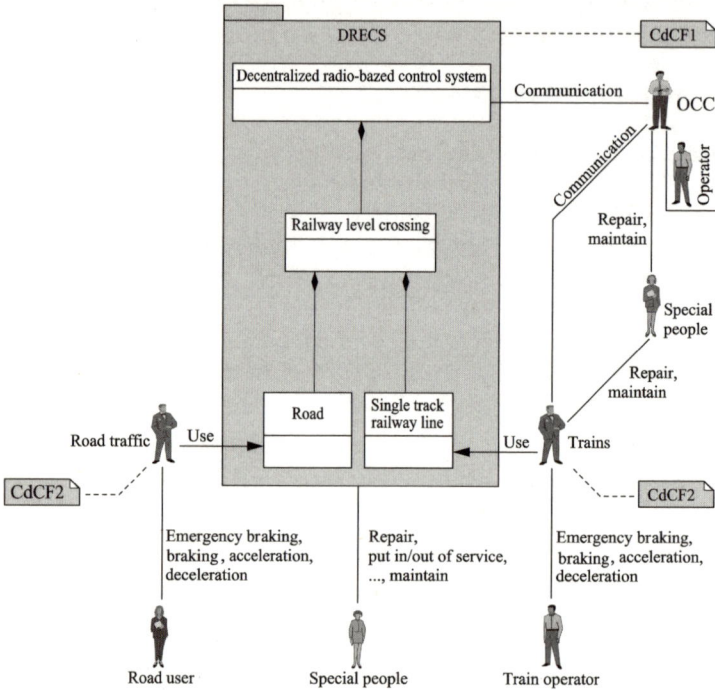

**Figure 2.3.** *Example of a model of the system in its environment*

Figure 2.3 shows a model of the control system of a railroad crossing. This system may command the intersection of at least one road with a railway track. It interacts with various actors, namely an OCC, users of the road (truck and car drivers, etc.), users of the railroad and operators in charge of the use and/or of the maintenance.

We have chosen to produce a diagram that models a decentralized management system for a level crossing composed of a railroad track and a road, using a decentralized communication system (DRBCS).

The main point of Figure 2.3 is to identify the actors interacting with the management system, either for their use of the road or of the train, the OCC and especially the maintenance operators and other operators (the model also identifies the "special people").

Identifying all the actors of the system is vital; if this stage is neglected, some actions may be omitted – such as maintenance operations – and it is possible to overlook perturbations or malfunctions. A classic example[4] is that of a WiFi network and how its efficiency can be related to the density of the networks around the system. We will not discuss the treatment of the human factor further in the following section as, although central, it does not concern the software-based critical packages.

## 2.3 System

This section will clarify the vocabulary for the implementation of software-based equipment. First of all, it should be noted that a software implementation is directly linked to its equipment and that without the hardware architecture, there can be no software. Indeed, the validation of software (see Chapter 6) requires hardware architecture and the results are only applicable to that particular hardware architecture. This is why the first definitions are those referring to the notion of system and software-based system.

DEFINITION 2.1 (SYSTEM).– *A system is an aggregate of interacting elements, organized so as to reach one or more pre-stated results.*

The "organized" side of definition 2.1 can be understood as exemplified in Figure 2.4.

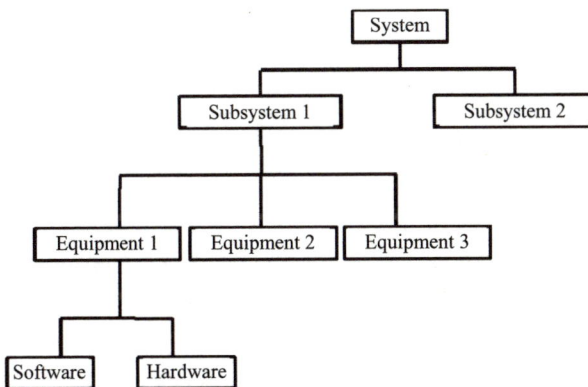

**Figure 2.4.** *From system to software*

4 The use of "open" networks such as WiFi carries inherent difficulties such as the densification of the networks (the number of networks is booming) and/or perturbations caused by annexed equipment. N.B. Open networks have not been treated in terms of dependability until recently as regards aspects of privacy, intrusions, etc., which are covered by the term "security".

Figure 2.4 puts forward a hierarchical vision of the system. A system performs several functions. It can be broken down into several subsystems, each performing functions which are sub-functions of the system's functions. As for the system, this vision must be complemented by a vision of the interactions between the functions such as those shown in figure 2.5.

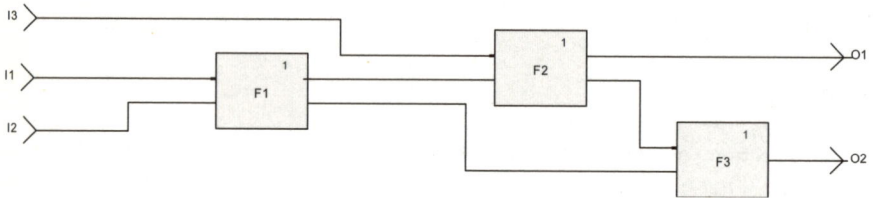

**Figure 2.5.** *Example of decomposition*

A subsystem supports several functions which in turn can be broken down into various pieces of equipment. The equipment is not a functional element in itself, but it must be completed by other equipment in order to perform a function.

DEFINITION 2.2 (SOFTWARE-BASED SYSTEM).– *Some elements from the system may be for part of or for the entirety of the software.*

In this chapter we have shown that a programmable computer-based system is a complex object, which should be analyzed and broken down accordingly.

## 2.4. Software implementation

### 2.4.1. *What is software?*

In this chapter, the term *software* is an aggregate of calculus/treatment elements, which are executed on the hardware architecture so that the resulting tool is able to perform tasks associated with some equipment (see Figure 2.4).

From here on, we shall focus on the aspects of the software, hence the need to define software; see definition 2.3.

DEFINITION 2.3 (SOFTWARE [AFN 97]).– *Set of programs, processes, rules and sometimes documentation relative to the functioning of a tool for treating information.*

Definition 2.3 does not distinguish between the means (methods, processes, tools, etc.) to perform the software implementation, the products resulting from it (documents, analysis results, models, sources, test scenarios, test results, specific tools, etc.) and the software implementation itself.

This definition is generally associated with the notion of software implementation. In turn, the notion of software is associated with that of an executable.

### 2.4.2. *Types of software*

Definition 2.3 highlights the notion of software, but it should be noted that there are several types of software:

– operational software: this refers to any software delivered to an external client within the framework of a program or product. Bank tests are part of this category;

– demonstrator: a demonstrator is used by an external client in order to refine its expression of need and measure the level of potential service. These types of software are not designed for operational usage;

– development tool: this accounts for an internal type of software, not delivered to an external client, designed to help development in the broad sense of the term (editor, compiler tool chain, etc.), including the tests and integration;

– model: this refers to software for internal study not delivered to external clients, used to verify a theory, and algorithm or the feasibility of a technique (e.g. by simulation), without any objectives, results or completeness.

### 2.4.3. *Software implementation in its context*

It is important to consider a given software implementation as a set of components (see definition 2.4) that interact to perform the treatment of given information. A component can thus be a program, a library, a trade component (COTS[5]), etc.

DEFINITION 2.4 (COMPONENT). – *A component is a software element that performs a number of given services, respecting a set of given requirements; it has definite interfaces and is treated as an element in itself in configuration management.*

---

5 COTS stands for Commercial Off-The-Shelf, COTS are commercially available products, which can be bought as they are (without specifications, V&V elements).

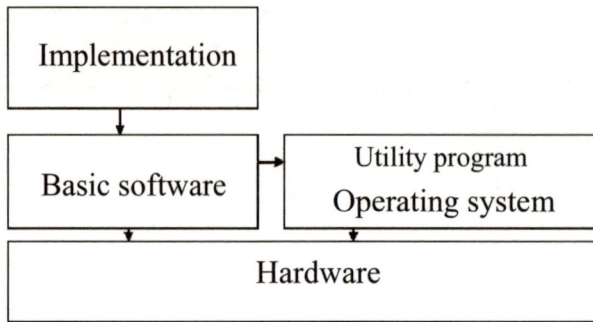

**Figure 2.6.** *An example of software implementation in its environment*

As shown in Figure 2.6, a software implementation generally uses an abstraction of the hardware architecture and of its operating system through a low layer called "basic software". Basic software generally writes in the low-level programming languages such as the assembler and/or the C. It not only allows us to encompass the services rendered by the operating system and by its utility programs, but it also gives a more or less direct access to material resources.

If the software implementation is associated with a high level of safety, then the lower layers (basic software, utility programs and operating systems) will themselves be associated with objectives related to safety. The safety objective of the lower layers will depend on the hardware architecture in place (mono processor, 2oo2, nOOm, etc.) and on the concepts of safety implemented. In [BOU 10, BOU 11], we put forward real examples of dependable architectures. In this book we shall suppose that the safety analyses have been performed and that a level of safety has been allocated to all software applications (including the lower layer).

## 2.5. Conclusion

This chapter has given us the opportunity to show the complexity underlying programmable computer-based systems. It has allowed us to understand the hierarchy involved in going from the system to the software via the subsystems, equipment and computer material.

In Chapter 4, we will show how to tackle the dependability of a system by mastering the electronics component of it (Chapter 5) and the software component (Chapter 6).

## 2.6. Bibliography

[AFN 90] NF F 71-013 Installation fixes et matériel roulant ferroviaires, Informatique, Sûreté de fonctionnement des logiciels – Méthodes appropriées aux analyses de sécurité des logiciels, December 1990.

[BOU 10] BOULANGER J.-L. (ed.), *Safety of Computer Architectures*, ISTE, London and John Wiley and Sons, New York, 2010.

[BOU 11] BOULANGER J.-L. (ed.), *Sécurisation des architectures informatiques industrielles*, Hermes-Lavoisier, Paris, 2011.

[ISO 08] ISO, ISO 9001: 2008, Systèmes de management de la qualité – Exigence, December 2008.

[NAU 69] NAUR P., RANDELL B. (eds), *Software Engineering: A Report on a Conference sponsored by NATO Science Committee*, NATO, 1969.

[SOM 07] SOMMERVILLE I., *Software Engineering*, version 8, 2007.

## 2.7. Glossary

COTS:           Commercial Off-The-Shelf

ISO[6]:           International Organization for Standardization

NATO:           North Atlantic Treaty Organization

nOOm:           n Out Of m architecture

OCC:            Operation Control Center

PAING:          *Poste   Aiguillage   Informatisé   –   Nouvelle   Génération* (Computerized Signaling Station – New Generation)

PIPC:           *Poste Informatique Petite Capacité* (Small Capacity IT Center)

PMI:            *Poste de Manœuvre Informatisé* (Computerized Work Station)

TCO:            *Tableau de Contrôle Optique* (Optical Control Table)

---

6 See http://www.iso.org.

# 3

# Certifiable Systems

Some systems need an authorization for operation and these authorizations can be based on a certification. However, the certification can be done independently on a used specific. When you need to demonstrate that your product fullfills a standard, certification can be mandatory. A certification is based on an independent assessment and needs all the elements (documents, electronic files,analysis, tests, etc.) resulting from the product realization to be auditable and analyzable by an external team.

## 3.1. Introduction

Dependable critical systems are characterized by the fact that a failure can produce serious consequences as much on people's lives as on the economy and/or the environment. In many fields, including the railway sector, standards make safety demonstrations mandatory for the systems in question. These standards are complemented by international (in the nuclear and spatial domain for example), European (in the railway domain for example) and/or national law and regulations (in manufacturing) which describe the rules to respect.

These standards advise on a separation between roles and responsibilities. One team should be in charge of the implementation of the system (development, verification and validation), whereas another team is assigned to the demonstration of the system's safety (safety checks, analysis of the safety records, completeness analyses). A third team, independent of the first two, assesses the level of safety actually achieved. This evaluation can (but does not have to) be formally recognized in the form of a certificate.

In this chapter we shall first describe the normative concept in some industrial sectors (robotic, aeronautical, railway, automotive, nuclear) and we shall present the aspects of assessment and certification.

## 3.2. Normative context

In this section we shall propose some normative contexts. The aim is not to present an exhaustive list of standards and their inter-relations but to give some landmarks which will be useful later.

[BAU 10, BAU 11] approaches standards by categorizing them by sector (aeronautical, automotive, railway, nuclear, spatial, and PLC[1]-based systems) and by comparing them through their similarities and differences.

### 3.2.1. *Generic standards*

#### 3.2.1.1. *Presentation*

Electrical/computer systems have been used for years to perform functions related to safety in most industrial sectors. The IEC 61508 ([IEC 08]) standard shows a generic approach towards all activities involved in the lifecycle of electrical/electronic/programmed electronics (E/E/PE) safety systems used to perform safety-related functions. This standard was edited by the International Electrotechnical Commission[2] (IEC) which is the international organization in charge of normalizing the fields of electricity, electronics and related technologies.

The IEC 61508 ([IEC 08]) standard proposes a globalized approach to safety in the sense of innocuousness[3], comparable to ISO[4] 9000 [ISO 08] in terms of quality.

The IEC 61508 standard ([IEC 08]) is subdivided into seven categories:

– IEC 61508-1, General objectives;

– IEC 61508-2, Objectives for electrical/electronic/programmed electronic systems involved in safety;

– IEC 61508-3, Objectives for software;

– IEC 61508-4, Definitions and abbreviations;

– IEC 61508-5, Examples of methods to determine levels of integrity and safety;

---

1 PLC stands for Programmable Logic Controller.

2 To learn more about it, visit: http://www.iec.ch.

3 The IEC 61508 standard ([IEC 08]) does not cover aspects of security, privacy and/or integrity related to the implementation of measures preventing unauthorized persons from damaging and/or affecting the security achieved by the E/E/PE appointed to safety issues. This includes managing networks so as to avoid intrusions.

4 http://www.iso.org/iso/home.html.

– IEC 61508-6, Directives for applying IEC 61508-2 and IEC 61508-3;

– IEC 61508-7, Holistic view of the measures and techniques.

The IEC 61508 standard ([IEC 08]) is coherent with the convergence observed among the various industrial sectors (aeronautical, nuclear, railway, automotive, manufacturing, etc.). Its content is complex and unusual enough for the reader to need guidance [ISA 05] or [SMI 07].

In most cases, safety is obtained through several systems with divergent technologies (mechanical, hydraulic, pneumatic, electrical, electronic, programmed electronic). The safety strategy must take into account all the elements contributing to safety. The IEC 61508 standard ([IEC 08]) provides a framework that can then be applied to safety-related systems based on other types of technology, before more specifically addressing computer-based systems.

Due to the great variety of applications of E/E/PE, as well as the various degrees of complexity involved, the exact nature of the safety measures to be implemented depends on factors specific to their application. This is why, in the IEC 61508 standard ([IEC 08]), there are no general rules but rather recommendations as to the methods of analysis to be used.

### 3.2.1.2. *Safety levels*

The IEC 61508 standard ([IEC 08]) defines the notion of SIL (safety integrity level). SIL allows the safety level of a given system to be quantified. Level of safety integrity 1 (SIL1) is the lowest level whereas Level 4 (SIL4) is the highest.

All four SIL levels for a given safety function are characterized by the impacts of its failures:

– SIL 4: catastrophic impacts (highest level);

– SIL 3: impacts on the community;

– SIL 2: major protection of the implant and of the production or risk of injury for the employees;

– SIL 1: minor protection of the implant and of the production (lowest level).

This standard details the requirements to be met for each SIL. The requirements are more stringent for higher levels so as to guarantee a lower probability of dangerous failure.

The SIL allows prescriptions for the safety functions to be specified and allocated to E/E/PE systems.

### 3.2.1.3. *Daughter standards*

For several years now, the IEC 61508 standard ([IEC 08]) has given rise to daughter (secondary) standards or variations which cover different fields as shown in Figure 3.1.

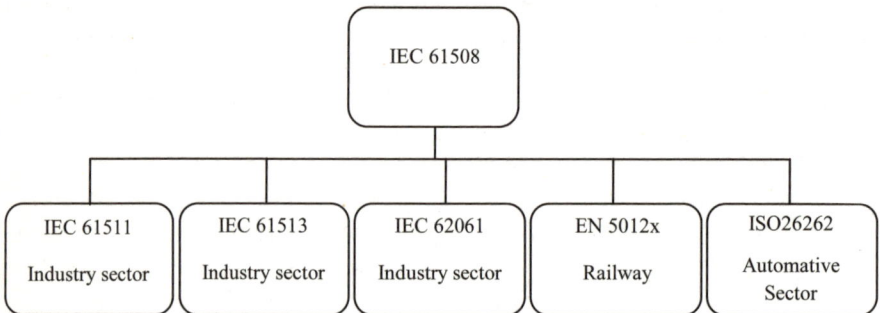

**Figure 3.1.** *IEC 61508 and its daughter standards*[5]

The industry standard IEC 61511 "Continuous Processes" ([IEC 05]) limits the field of application of the IEC 61508 standard ([IEC 08]) to the traditional context of continuous processes. In so doing, it is more precise.

It is to be noted that, as far as programming languages are concerned, the IEC 61511 standard ([IEC 05]) only considers the most common programming languages which are highlighted in IEC 61131-3[6] ([IEC 03]). This, however, does not exclude the use of the languages ADA or C++ for programming, but, in those cases, IEC 61508 ([IEC 08]) must be referred to as it gives a framework for those particular languages.

The CENELEC standards (EN 50126 [CEN 00], EN 50129 [CEN 03], and EN 50128 [CEN 01a, CEN 11]) will be discussed in section 3.2.2. The daughter

---

5 Figure 3.1 highlights a link between standard IEC 61508 and IEC 61513 standards, however this is incorrect as nuclear standards existed before the IEC 61508 standard and the link is present in name only.

6 IEC 61131 is an IEC standard for programmable controllers (PLC). Part 3 refers to languages for programmable controllers.

standard for the automotive sector ISO 26262 ([ISO 09]) will be described in section 3.2.3.

### 3.2.2. *Railway sector*

This section focuses on our return as assessors in the railway sector (urban and main railway lines). An evaluation of the implementation of railway standards was made in [BOU 07].

#### 3.2.2.1. *History from CENELEC to IEC*

Electronic/computer systems have been used for years to perform functions related to safety in most industry sectors. The IEC 61508 standard [IEC 08] gives a generic approach to all the activities involved in the safety lifecycle of electrical/electronic/programmed electronic systems (E/E/PES) used in the performance of those safety-related functions.

In most cases, safety is obtained through several systems with divergent technologies (mechanical, hydraulic, pneumatic, electrical, electronic, programmed electronic). The safety strategy must take into account all the elements contributing to safety. The IEC 61508 standard ([IEC 08]) provides a framework that can then be applied to systems related to safety, based on other types of technology, before addressing more specifically computer-based systems.

Due to the great variety of applications of E/E/PE as well as the various degrees of complexity involved, the exact nature of the safety measures to be implemented depends on factors specific to their application. This is why, in IEC 61508 ([IEC 08]), there are no general rules but rather recommendations as to the methods of analysis to be used.

#### 3.2.2.2. *CENELEC standards*

Projects in the railway sector are dictated by texts (decrees, laws, etc.) and a set of standards (CENELEC[7] standards EN 50126 [CEN 00], EN 50129 [CEN 03], and EN 50128 [CEN 01a, CEN 11]) designed to define and achieve RAMS objectives (reliability, availability, maintainability and safety). The three standards cover aspects of safety in the sense of innocuousness of the system down to the hardware and/or software.

---

7 European Committee for Electrotechnical Standardization; see www.cenelec.eu.

### 3.2.2.3. *Presentation*

As shown in Figure 3.2, the railway standard CENELEC EN 5012x is a variation of the generic IEC 61508 standard ([IEC 08]) that takes into account the specificities of the railway sector and of successful experiments (SACEM, TVM, SAET-METEOR, etc.).

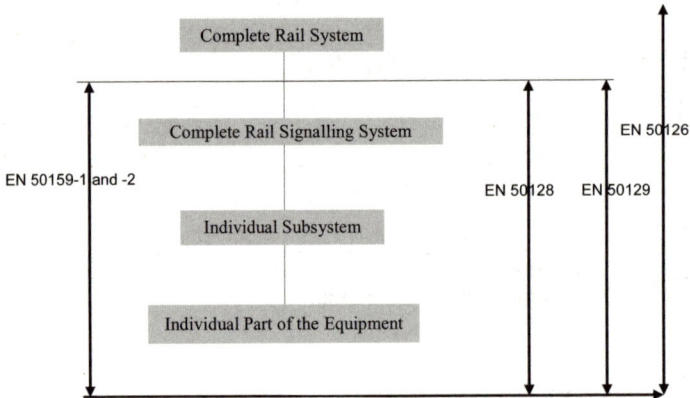

**Figure 3.2.** *Standards applicable to rail systems*

Railway standard CENELEC EN 5012x applies to "urban" railway applications (metro, suburban express train, etc.) much more than for conventional applications (high speed line, ordinary train lines, freight).

One of the restrictions of this railway standard has to do with the field of implementation of the CENELEC EN 50128 [CEN 01a, CEN 11] and CENELEC EN 50129 standards [CEN 03] as these are normally restricted to signaling subsystems (see Figure 3.3). For hardware architecture used in other subsystems, older standards can be found which are still applicable (NF F 00-101, etc.). Work is being done within certified bodies to extend and generalize these two standards to the entire rail system.

In the railway sector, the normative repository is made up of the following standards:

– the CENELEC EN 50126 standard [CEN 00] describes measures to be implemented in order to define and establish reliability, availability, maintainability and safety (RAMS);

– the CENELEC EN 50128 standard [CEN 01a, CEN 11] describes actions to be taken so as to show the safety of the software;

– the CENELEC EN 50129 standard [CEN 03] describes the structure of the safety record as well as the measures to undertake in order to achieve hardware safety.

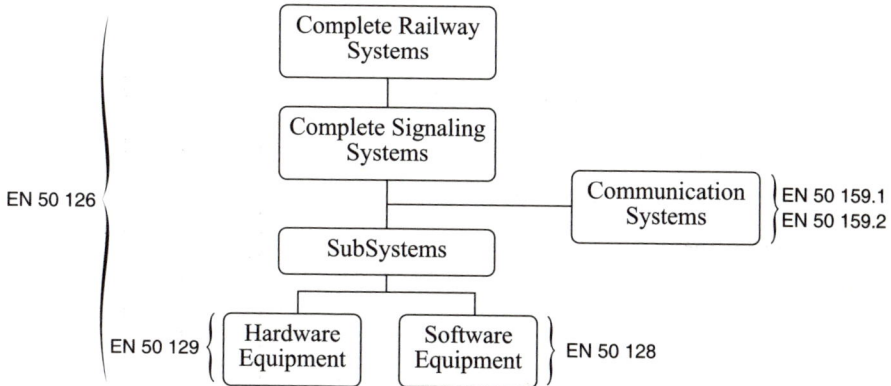

**Figure 3.3.** *Organization of standards applicable to rail systems*

So, the CENELEC EN 50129 standard is more oriented towards achieving hardware safety. In [BOU 10] the principles of hardware safety were presented, whereas in [BOU 09, BOU 11], examples of dependable hardware were given.

The CENELEC EN 50128 standard ([CEN 01a, CEN 11]) allows the management of software application safety. In [BOU 11a] techniques to enforce safety of a software application have been presented.

These standards are complemented by one related to "transmission" aspects. The 2001 version of CENELEC EN 50159 was broken down into two parts: the first one EN 50159-1 [CEN 01b] dedicated to closed networks, and the second EN 50159-2 [CEN 01c] for open networks. The new version CENELEC EN 50159 [CEN 11a] replaces the previous one and covers both open and closed networks. Thus, CENELEC EN 50159 allows us to tackle apsects of security in the sense of privacy.

On top of the standards referring to the equipment's architecture (software + hardware), there are additional standards to regulate its interactions with the environment. One such standard is CENELEC EN 50121 [CEN 06] which addresses problems of electromagnetic compatibility (EMC), both in terms of the rail system toward the outside world, as well as from the point of view of the environment toward the rail system (train and built-in equipment).

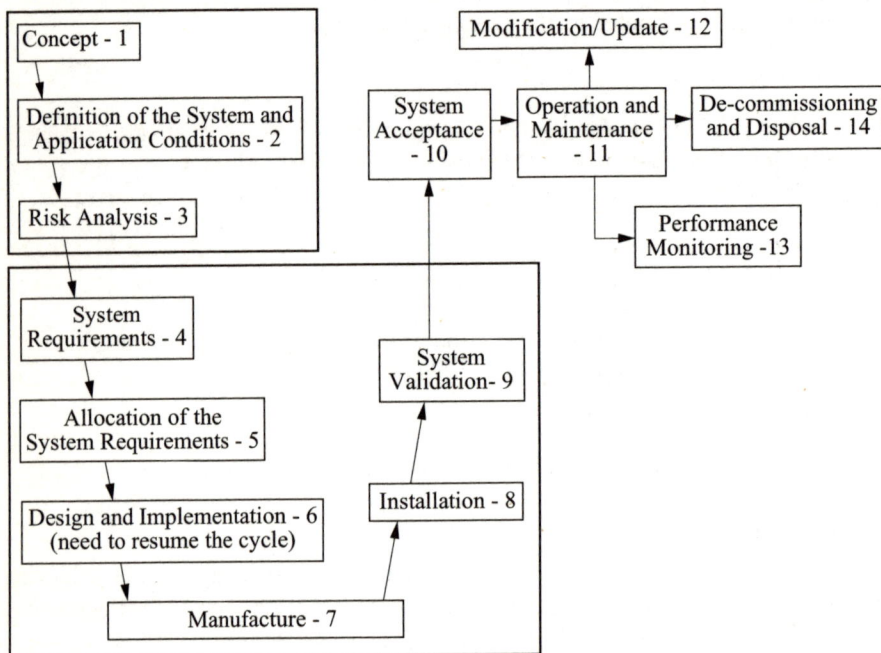

**Figure 3.4.** *Safety lifecycle*

Figure 3.4 places the safety lifecycle in the context of the railway sector. The cycle is composed of three stages:

– preliminary analysis (stages 1, 2 and 3);

– implementation of the system, subsystem and/or equipment (stages 4 to 9);

– installation and usage of the system (stages 10 to 14).

According to the standard CENELEC EN 50129 [CEN 03], the requirements can be separated into those relative to a system's safety and those not concerned with safety. CENELEC EN 50126 [CEN 00] proposes the following steps (we shall only consider the aspects relative to the implementation process) to specify and establish the safety of a given system:

– definition of the system and the conditions for its application: mission profile, description of the system, exploitation and maintenance strategy, and identification of the constraints generated by the existing system or line;

– risk analysis;

– identification of the system's requirements: analysis of the demands, specification of the system and its environment, definition of the demonstration and acceptation criteria for the system;

– allocation of the  system's requirements: specification of the requirements for the subsystems, equipment and/or components and definition of the acceptation criteria for these elements.

Design and implementation: realization of the design, development and verification, and validation process.

### 3.2.2.4. *Implementation*

The purpose of the CENELEC referential is to:

1. provide standards common for the whole of Europe in order to favor the opening of markets for rail components, interoperability, interchangeability and "cross-acceptance" [CEN 07a] of rail components;

2. respond to the specificities of the railway sector.

Safety, when taken as a component of dependability, is obtained through the implementation of concepts, methods and tools throughout the lifecycle. A safety assessment requires analysis of the system's failures. The seriousness of the potential consequences must be identified and quantified and, whenever possible, the expected frequency of those failures should be calculated.

Among the actions to reduce risk so as to reach an acceptable level, the CENELEC EN 50129 standard [CEN 03] proposes allocating safety objectives (see Chapter 4) to the various functions of the system and its subsystems. The SIL is defined as one of the discrete levels specifying the safety integrity requirements of the safety functions allocated to the safety system.

The CENELEC EN 50128 standard [CEN 01a] is more appropriate to aspects of software development for the railway sector. When it comes to software, the SSIL

(Software SIL) allows us to define the level of severity from 0 (not dangerous) to 4 (critical).

Safety requirements have always been a factor within complex systems (rail transport, air transport, nuclear power plants). However, nowadays, contractual obligations on performance, for example, have brought manufacturers, in the railway sector, to the total control of parameters acting on reliability, availability and maintainability (RAM). So the choice of standards now falls to the designer and/or manufacturer.

### 3.2.2.5. *Safety vs availability*

Standards describe the processes, methods and tools necessary to meet and establish the level of SIL required. These obligations are imposed on the means and are added to obligations on quantitative and/or qualitative results.

So far, we have emphasized the safety aspect (in the sense of innocuousness) as it remains the main aspect of dependability and its analysis is mandatory. There are few or no requirements for reliability from the normative framework.

For systems with a critical SIL level (3 and 4), the principles of enforcing safety are in direct contradiction with the availability of the system. As a matter of example, in the railway sector, the state of safety, which corresponds to the "train stop", has a strong impact on the global availability of the system. On the other hand, obligations surrounding the manufacturing of critical systems allows improvements in the reliability of the software.

As to software applications of a non-critical (SIL 0) and low criticality (SIL 1 and SIL 2) type, the manufacturing process of the software is less constrained (both when it comes to the choice of programming languages and tools regarding the processes), which induces a lower software quality and entails a low reliability of the software. More precisely, for non-critical systems or systems with a low critical level, manufacturers can use products called "commercial off-the-shelf" (COTS). As quality control has a direct impact on reliability and availability, COTS remains a topical matter.

### 3.2.2.6. *Future*

As shown for example by the recent evolution of CENELEC EN 50128 [CEN 11], the CENELEC standard is in the process of a complete restructuring. Work is being done among certified bodies to extend and generalize standards EN 50128 and EN 50129 across the whole rail system through a new version of the CENELEC EN 50126 standard.

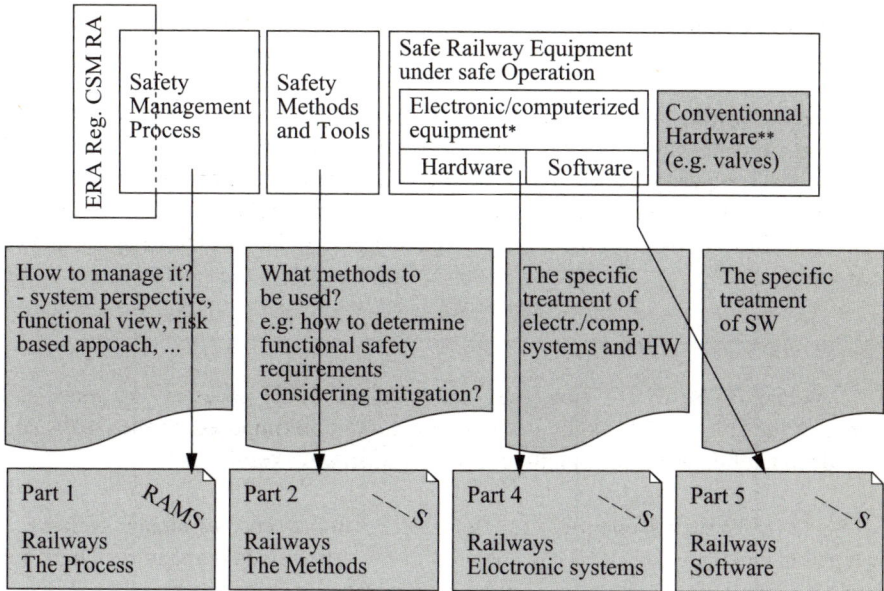

**Figure 3.5.** *Future version of CENELEC EN 50126 standard*

### 3.2.2.7. Software safety

The CENELEC EN 50128 standard [CEN 01a, CEN 11] provides the procedures and technical prescriptions applicable to the development of the programmed electronic systems used in command and protection rail applications. The CENELEC EN 50128 standard does not usually apply to all software applications of the railway sector. CENELEC EN 50155 [AFN 07] on the other hand applies to all embedded applications on a train.

### 3.2.2.7.1. CENELEC EN 50155

The EN 50155 standard [AFN 07] is applicable to all electronic equipment[8] present on trains. This standard does not concern software applications exclusively; nevertheless, section 5.3 of the standard describes the rules that apply to software aspects in detail.

---

8 These can be electronic equipment for command (TCMS, TCN, etc.), traction, regulation, protection or supply, etc.

The first requirement has to do with the mandatory observance of the ISO 9001:2008 standard [ISO 08]. The second requirement regards the implementation of management and set-up procedures that cover the developed software applications but also the tools used for its development and maintenance.

The third requirement deals with defining a lifecycle for the development of a software application, which must be well-structured and documented.

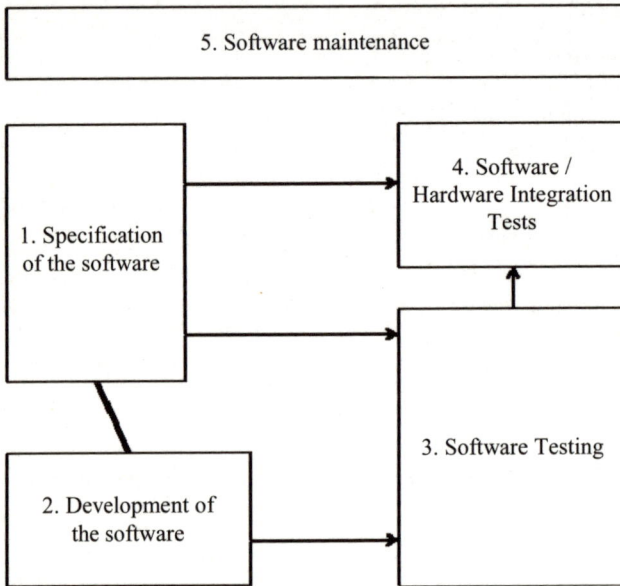

**Figure 3.6.** *Development cycle – EN 50155*

As shown in Figure 3.6, the development cycle proposed by standard EN 50155 [AFN 07] can be broken down into five stages:

– specifying the software;

– designing the software;

– testing the software;

– testing software/hardware integration;

– maintaining the software application.

For the specification phase, the EN 50155 standard [AFN 07] introduces the notion of "requirement" and the specification document without further explaining this notion of requirement.

As to the software development phase, section 5.3.2 of EN 50155 [AFN 07] indicates that a modular approach is needed (software is decomposed into small-sized components, hence the limitation on their size and number of interfaces). It further alludes to the need for a structured approach (procedures, document list, standard design, uniformity, etc.) and for structured methods (logical/functional diagram-blocks, sequential diagrams, diagrams for the flux of data, truth/decision tables). Furthermore, when it comes to the code, it is necessary to choose a programming language that facilitates verification (that requires minimum effort to understand). The design method and programming language must make the analysis of the code easy and the behavior of the program must be deductible from the analyses of the code.

As far as the testing of the software is concerned, the EN 50155 standard [AFN 07] recommends the analysis of boundary values, equivalence classes, as well as the partitioning of inputs. The selection of the cases to be tested can be supported by a model that simulates the process.

The maintenance phase is a key point as it is crucial to guarantee that any modification, addition of functions or adaptation of the software application, for a different type of train for example, does not compromise safety. The management of the maintenance phase must be fully defined and documented. One of the challenges of the maintenance phase is to manage the evolution of several parallel versions as well as the installation in the field (the trains being of different types and parked in different places).

The EN 50155 standard [AFN 01] advocates that all the elements produced should be registered in a format that allows later analysis. This standard gives the following list as an indication of the documents that need to be produced:

– specification of the requirements for the software with a presentation of the coverage of the system's requirements;

– description of the hardware and of the software application design enables the needs expressed in the software requirement specifications to be met;

– for each of the software components, a description of the performance characteristics (list of inputs and outputs, etc.) must be provided, as well as the source code and the description of trial results;

– all the data from the software application (global variables and constants) must be described in the so-called "data dictionary";

– description of the software application lay-out in storage;

– description and referencing (name, version) of the tools used to develop the software application and to produce the executable;

– description of the integration trials and the results associated with them.

In § 8.2, EN 50155 [AFN 01] makes standards CENELEC EN 50126 [CEN 00], EN 50129 [CEN 03], and EN 50128 [CEN 01a] applicable without clearly stating how or on which perimeter.

### 3.2.2.7.2. CENELEC EN 50128

The CENELEC EN 50128 standard [CEN 01a, CEN 11] applies to applications related and unrelated to safety[9] and exclusively to software and the interaction between the software application and the system as a whole.

This standard advocates the implementation of a cycle V whose coverage goes from the software specification to the tests on the software. One of the particularities of standard EN 50128 lies in the necessity to show the deployment of means. This is why it is defined as a standard of means.

The CENELEC EN 50128 standard [CEN 01a, CEN 11] explicitly introduces the notion of assessment. As shown in [BOU 06b, BOU 07], in the case of software applications, the assessment of the software consists of demonstrating that the software application meets the safety objectives attributed to it.

Assessing the level of safety for a given software application is based on:

– the development of an audit: is the quality control effective on the project?;

– the review of the plans (software quality assurance, test(s), V&V, etc.);

– the review of the produced elements (documents, source, generation chain of the executable, production process for the data, scenarios and test results, safety analyses (e.g. SEEA[10]), etc.);

– the formalization of the potential caveats and non-conformities;

– the formalization of a warning in the form of an assessment report.

The CENELEC EN 50128 standard has an IEC application in the IEC 62279. The last version of the IEC 62279 is a copy of the

---

9 This is why standard EN 50128 [CEN 01a, CEN 11] introduces the SSIL0 which is related to software applications without any safety requirements.

10 SEEA stands for software error effect analysis.

EN 50128:2001standard but for the new version of EN 50128:2011, the associated version of the 62279 introduces some improvements (including in the section that describes the tool qualification and some changes to appendix A).

### 3.2.3. *Automotive sector*

As highlighted in Figure 3.1, the ISO 26262 standard [ISO 11] is a variation of the generic standard IEC 61508 [IEC 08] which takes into account the particularities of the automotive sector.

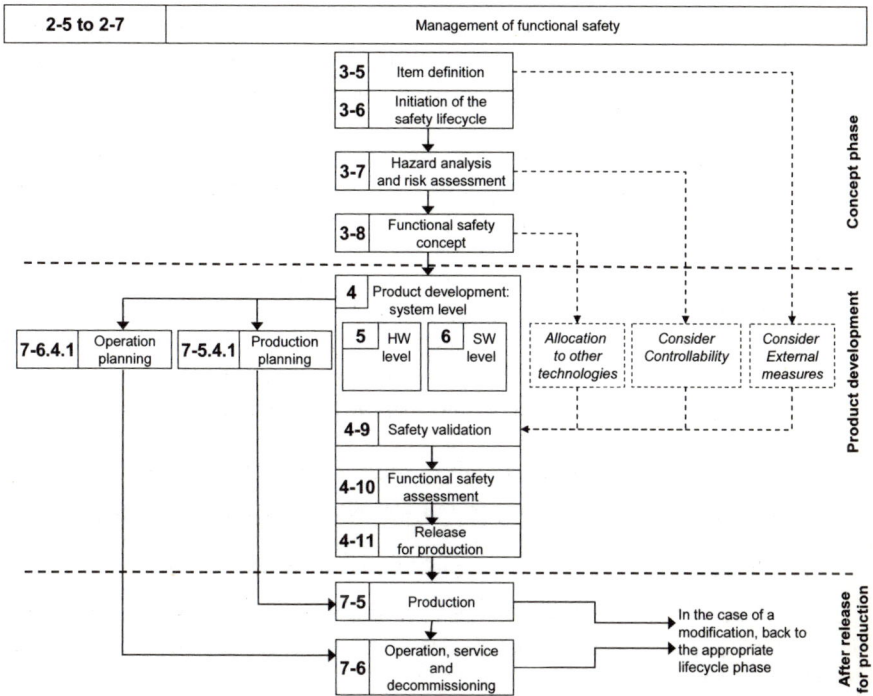

**Figure 3.7.** *Safety lifecycle*

The ISO 26262 standard [ISO 11] proposes a safety control process that goes from defining to managing withdrawal as shown in Figure 3.7.

Part 5 of ISO 26262 [ISO 09] is devoted to controlling the hardware component of the equipment; see Figure 3.8.

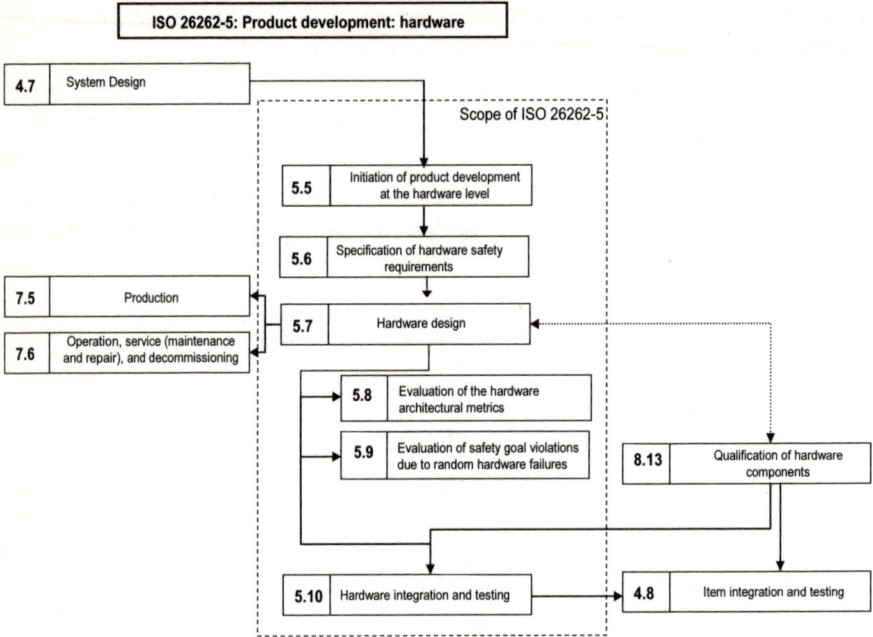

**Figure 3.8.** *Achieving hardware safety*

In [LIA 08], an appreciation is given of the impacts of implementing ISO 26262 [ISO 09] on the processes (defining the ASIL[11], managing safety, etc.). Appendix D of part 5 introduces the various techniques for securement. For some examples of automotive architectures, consult [BOU 11].

Part 6 of ISO 26262 ([ISO 11]), on the other hand, concerns software safety control.

As to the implementation of the ISO26262 standard [ISO 11], the various industries in the automotive sector have started to measure the impact of its implementation on processes currently in use; see [LIA 08] and [AST 10].

---

11 ASIL stands for automotive SIL, the different values are A, B, C and D. The highest level is D and the lowest level is A, but QM (quality management) also exists.

### 3.2.4. *Aviation sector*

3.2.4.1. *Presentation*

The aeronautical field possesses a repository (FAR[12] and JAR[13]) which defines the targets at the level of the system (10–9/h), the methods and the means to be deployed. This repository is followed by a common "practical" repository made up of a set of ARP (aerospace recommended practices) which includes:

– ARP 47.54 [ARP 4754] entitled "Certification Considerations for Highly-Integrated or Complex Aircraft", describes the methods to achieve safety analyses for aviation systems;

– ARP 47.61 [ARP 4761] "Guidelines and Methods for Conducting the Safety Assessment Process of Civil Airborne System and Equipment" delineates certification for complex aviation systems.

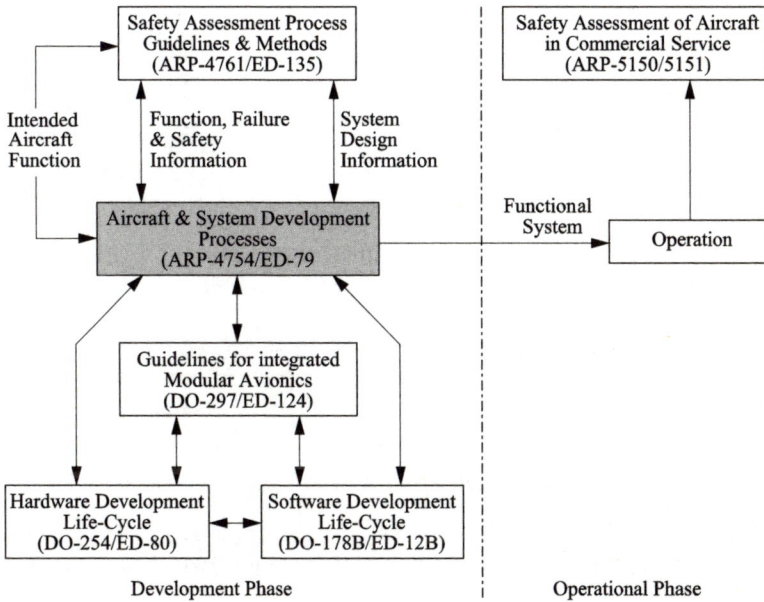

**Figure 3.9.** *Development cycle in aeronautics*

The methods for safety assessment are detailed and depicted in document ARP-4761.

---

12 FAR stands for Federal Aviation Regulation.
13 JAR stands for Joint Aviation Requirement.

From a normative point of view, the aviation sector refers to the DO 178B standard [ARI 11] for software aspects and to DO 254 [ARI 00] for electronics questions.

Figure 3.9 illustrates the organization and articulation between the different standards applicable to the system, software etc. in the aeronautics field.

More precisely, the DO178 standard exemplifies the direction lines to be followed in order to produce codes for aircraft applications. The DO 178 standard therefore is a complement to the standard ARP 47.54 (see Figure 3.9) which specifies the designing process for control steering and modification of the existing systems. The DO 178 standard was updated in 2011, hence version C.

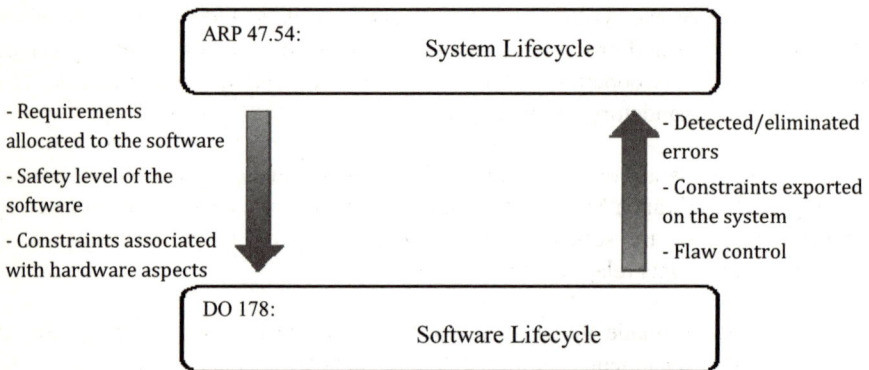

**Figure 3.10.** *Relations between system and software*

3.2.4.2. *Software safety level*

The aviation field defines the notion of DAL (design assurance level); these levels are as follows:

– Level A: catastrophic problem – flight or landing safety compromised – plane crash;

– Level B: major problem entailing serious damage and possible death of some of the occupants;

– Level C: serious problems causing malfunctioning in vital equipment of the apparel;

– Level D: problem with a potential to compromise the safety of the flight;

– Level E: problem with no effects on the safety of the flight.

The DO-178 standard puts forward an objective-oriented approach: the objectives are laid out, then a verification as to whether they have been met is performed. However, the precise methods to be used are not specified (nor is the development lifecycle).

### 3.2.5. *Aerospace*

In the aerospace sector, the European Space Agency (ESA) has published the PSS[14] (product assurance and safety standards) that has variations for all the elements of a system. This repository provides a general approach towards safety, the steps to conduct required studies, as well as the objectives to be met, etc.

This repository was written by the ESA and there is a study group working on improving the PSS and providing a repository endorsed by industrial entrepreneurs (ECSS: European Cooperation for Space Standardization). Nevertheless, this repository is not mandatory and it simply sets the basis for all projects.

ECSS standards are organized under three main branches, namely management and organization (branch M), quality, product assurance (branch Q) and engineering (branch E). These are supplemented by general aspects (such as the glossary) (branch S) and aspects related to the systems.

Each branch is made up of a first level of standards, and, if appropriate, of complementary documents (more detailed standards, guides facilitating the implementation of the first level of standards, or handbooks providing additional, non normative, information).

### 3.3. Conclusion

We can conclude that whatever the field, there is a repository of standards which proposes a scale allowing us to qualify the level of criticality of a system. This level of criticality defines the effort to be deployed. The safety control for a critical system is achieved through defining the process while relying mostly on testing.

As to the reliability aspect, for critical systems, reliability follows from quality control. As opposed to non-critical systems, which are actually COTS, there is currently a real issue in defining effective reliability, even though it is crucial information.

---

14 http://esapub.esrin.esa.it/pss/pss-cat1.htm.

## 3.4. Bibliography

[AFN 07] AFNOR, EN 50155, French standard, Applications Ferroviaires. Equipements électroniques utilisés sur le matériel roulant, October 2007.

[AHM 09] AHMED A., LANTERNIER B., "Trois méthodes pour déterminer le niveau de sécurité", *Revue Mesure*, no. 813, pp 26-27, 2009.

[AIC 93] American Institute of Chemical Engineers, *Guidelines for Safe Automation of Chemical Processes*, CCPS, New York, 1993,

[ARI 11] Software Considerations in Airborne Systems and Equipment Certification, published by ARINC, no. DO 178B, and EUROCAE, no. ED12, edition C, 2011.

[ARI 00] Design Assurance Guidance for Airborne Electronic Hardware, published by ARINC, no. DO254, and EUROCAE, no. ED80, 2000.

[ARP 96a] Certification Considerations for Highly-Integrated or Complex Systems, published by SAE, no. ARP4754, and EUROCAE, no. ED79, 1996.

[ARP 96b] Guidelines and Methods for Conducting the Safety Assessment Process on Civil Airborne Systems, published by SAE, no. ARP4761, EUROCAE, no. 135, 1996.

[AST 10] ASTRUC J-M., BECKER N., "Toward the application of ISO 26262 for real-life embedded mechatronic systems", *ERTS2*, 19-21 May 2010, Toulouse, France.

[BAR 90] BARANOWSKI F., Définition des objectifs de sécurité dans les transports terrestres. Technical Report 133, INRETS-ESTAS, 1990.

[BAU 10] BAUFRETON P., BLANQUART J.P., BOULANGER J.L., DELSENY H., DERRIEN J.C., GASSINO J., LADIER G., LEDINOT E., LEEMAN M., QUÉRÉ P., RICQUE B., "Multi-domain comparison of safety standards", *ERTS2*, 19-21 May 2010, Toulouse France

[BAU 11] BAUFRETON P., BLANQUART J.-P., BOULANGER J.-L., DELSENY H., DERRIEN J.-C., GASSINO J., LADIER G., LEDINOT E., LEEMAN M., QUERE P., RICQUE B. "Comparaison de normes de sécurité-innocuité de plusieurs domaines industriels", *Revue REE*, No 2 - 2011

[BIE 98] BIED-CHARRETON D., Sécurité intrinsèque et sécurité probabiliste dans les transports terrestres, Technical Report 31, INRETS - ESTAS, November 1998.

[BLA 08] BLAS A. and BOULANGER J.-L., "Comment Ameliorer les Méthodes d'Analyse de Risques et l'Allocation des THR, SIL et autres Objectifs de Sécurité", *LambdaMu 16, 16ème Congrès de Maîtrise des Risques et de Sûreté de Fonctionnement*, Avignon 6-10 October 2008.

[BOU 00] BOULANGER J.-L. and GALLARDO M., "Processus de validation basée sur la notion de propriété", *LamnbdaMu 12*, 28-30 March 2000.

[BOU 06a] BOULANGER J.-L., Expression et validation des propriétés de sécurité logique et physique pour les systèmes informatiques critique, May 2006, Compiègne University of Technology, France.

[BOU 06b] BOULANGER J.-L. and SCHÖN W., "Logiciel sûr et fiable : retours d'expérience", *Revue Génie Logiciel*, December 2006, No 79, pp 37-40.

[BOU 07] BOULANGER J.-L. and SCHÖN W., "Assessment of Safety Railway Application", *ESREL 2007*.

[BOU 08] GALLARDO M. and BOULANGER J.-L. "Poste de manoeuvre à enclenchement informatique: démonstration de la sécurité", *CIFA, Conférence Internationale Francophone d'Automatique*, Bucharest, Romania, November 2008.

[BOU 09] BOULANGER J.L. (ed.), *Sécurisation des architectures informatiques – exemples concrets*, Hermès-Lavoisier, Paris, 2009.

[BOU 09b] BOULANGER J.-L., "Le domaine ferroviaire, les produits et la certification"., *Journée "ligne produit" 15 October 2009*, Ecole des mines de Nantes, France.

[BOU 10] BOULANGER J.-L., "Sécurisation des systèmes mécatroniques. Partie 1", dossier BM 8070, *Revue technique de l'ingénieur*, November 2010.

[BOU 11] BOULANGER J.-L. (ed.), *Sécurisation des architectures informatiques industrielles*, Hermès-Lavoisier, Paris, 2011.

[BOU 11a] BOULANGER J.-L., "Sécurisation des systèmes mécatroniques. Partie 2", dossier BM 8071, *Revue technique de l'ingénieur*, April 2011.

[BOU 99] BOULANGER J.-L., DELEBARRE V. and NATKIN S., "Meteor: Validation de Spécification par modèle formel", *Revue RTS*, No 63, pp47-62, April-June 1999.

[CEN 00] CENELEC, NF EN 50126, Applications Ferroviaires. Spécification et démonstration de la fiabilité, de la disponibilité, de la maintenabilité et de la sécurité (FMDS), January 2000.

[CEN 01a] CENELEC, NF EN 50128, Applications Ferroviaires. Système de signalisation, de télécommunication et de traitement – Logiciel pour système de commande et de protection ferroviaire, July 2001.

[CEN 01b] CENELEC, EN 50159-1, European standard, Applications aux Chemins de fer : Systèmes de signalisation, de télécommunication et de traitement - Partie 1: communication de sécurité sur des systèmes de transmission fermés, March 2001.

[CEN 01c] CENELEC, EN 50159-2, European standard, Applications aux Chemins de fer: Systèmes de signalisation, de télécommunication et de traitement - Partie 2: communication de sécurité sur des systèmes de transmission ouverts, March 2001.

[CEN 03] CENELEC, NF EN 50129, European standard, Applications ferroviaires : systèmes de signalisation, de télécommunications et de traitement systèmes électroniques de sécurité pour la signalisation, 2003.

[CEN 06] CENELEC, EN 50121, Railway applications - Electromagnetic compatibility, 2006.

[CEN 07]  CENELEC, European standard, Railway Applications – Communication, Signalling and Processing systems – Application Guide for EN 50129 – Part 1: cross-acceptance, May 2007.

[CEN 11] CENELEC, EN 50128, European standard, Applications Ferroviaires. Système de signalisation, de télécommunication et de traitement – Logiciel pour système de commande et de protection ferroviaire, October 2011.

[CEN 11a] CENELEC, EN 50159, European standard, Applications aux Chemins de fer: Systèmes de signalisation, de télécommunication et de traitement - Communication de sécurité sur des systèmes de transmission, August 2011.

[DG 06] DG Énergie et Transport. Ertms - pour un trafic ferroviaire fluide et sûr. Technical report, Commission Européenne, 2006.

[GEF 98]  GEFFROY J.-L. and  MOTET G., *Sûreté de fonctionnement des systèmes informatiques*, InterEditions 1998.

[GOL 06]  GOLOUBEVA O.,  REBAUDENGO M,  SONZA REORDA M. and  VIOLANTE M., *Software Implemented Hardware Fault Tolerance*, Springer, 2006.

[GUL 04]  GULLAND W. G., "Methods of Determining Safety Integrity Level (SIL) Requirements - Pros and Cons", *Proceedings of the Safety-Critical Systems Symposium*, Birmingham (England), February 2004.

[HAD 95] HADJ-MABROUCK H., "La maitrise des risques dans le domaine des automatismes des systèmes de transport guidés: le problème de l'évaluation des analyses préliminaires de risqué", *Revue Recherche Transports Sécurité*, (49):101–112, December 1995.

[HAD 98] HADJ-MABROUCK H., STUPARU A., and BIED-CHARRETON D., "Exemple de typologie d'accidents dans le domaine des transports guides", *Revue générale des chemins de fers*, 1998.

[HAR 06]  HARTWIG K.,  GRIMM M.,  MEYER ZU HÖRSTE M., *Tool for the allocation of Safety Integrity Levels, Level Crossing*, Montreal, 2006

[IEC 91] IEC, IEC 1069: Mesure et commande dans les processus industriels - appréciation des propriétés d'un système en vue de son évolution, Technical report, 1991.

[IEC 98] IEC, IEC 61508: Sécurité fonctionnelle des systèmes électriques électroniques programmables relatifs à la sécurité, international standard, 1998.

[IEC 03]  IEC, IEC 61131-3, Programmable controllers – Part 3: Programming languages 2003

[IEC 05] NF EN 61511, European standard, Sécurité fonctionnelle Systèmes instrumentés de sécurité pour le secteur des industries de transformation, March 2005.

[ISA 05] ISA, Guide d'Interprétation et d'Application de la Norme IEC 61508 et des Normes Dérivees IEC 61511 (ISA S84.01) ET IEC 62061, April 2005.

[ISO 05] EN ISO/IEC 17025, Exigences générales concernant la compétence des laboratoires d'étalonnages et d'essais, 2005.

[ISO 08] ISO, ISO 9001:2008, Systèmes de management de la qualité - Exigence, December 2000.

[ISO 11] ISO, ISO/DIS26262, Road vehicles – Functional safety, 2011.

[LAP 92] LAPRIE J.C., AVIZIENIS A., KOPETZ H. (ed.), *Dependability: Basic Concepts and Terminology, Dependable Computing and Fault-Tolerant System*, vol. 5, Springer, New York, 1992.

[LIA 08] LIAIGRE D., "Impact de ISO 26262 sur l'état de l'art des concepts de sécurité automobiles actuels", *LambdaMu'08*, Avignon, October 2008.

[LIS 95] LIS, Laboratoire d'Ingénierie de la Sûreté de Fonctionnement. *Guide de la sûreté de fonctionnement*, Cépaduès, 1995.

[LEV 95] LEVESON N.G., *Safeware: System Safety and Computers*, Addison Wesley, Nenlo Park, CA. First ed., 1995.

[PAP 10] PAPADOPOULOS Y., WALKER M., REISER M.-O., WEBER M., CHEN D., TÖRNGREN M., SERVAT D., ABELE A., STAPPERT F., LONN H., BERNTSSON L., JOHANSSON R., TAGLIABO F., TORCHIARO S, SANDBERG A., "Automatic Allocation of Safety Integrity Levels", *Workshop CARS, Congrès EDCC 2010*.

[RSS 07] RAIL SAFETY AND STANDARDS BOARD, *Engineering Safety Management (The Yellow Book) – Fundamentals and Guidance*, Volumes 1 and 2, Issue 4, 2007.

[SIM 06] SIMON C., SALLAK M. and AUBRY J.F., "Allocation de SIL parAgregation d'Avis d'Experts SIL Allocation by Aggregation of Expert Opinions", *LambdaMu 15*, 2006

[SMI 07] SMITH D.J. and KENNETH G. L. Simpson, *Functional Safety, A Straightforward Guide to Applying IEC 61508 and Related Standards*, second edition, Elsevier, 2007.

[STR 06] STRMTG, Mission de l'Expert ou Organisme Qualifié Agrée (EOQA) pour l'évaluation de la sécurité des projets, version 1 du 27/03/2006.

[VIL 88] VILLEMEUR A., *Sûreté de fonctionnement des systèmes industriels*, Eyrolles, Paris, 1988.

## 3.5. Glossary

API:        Application Programming Interface

ARP:        Aerospace Recommended Practice

ASIL:       Automotive Safety Integrity Level

CENELEC[15]: *Comité Européen de Normalisation Électrotechnique* (European Committe for Electrotechnical Standardization)

---

15 See www.cenelec.eu.

COFRAC:    *COmité FRançais d'Accréditation* (French Accreditation Committee)

COTS:    Commercial off-the-shelf

DAL:    Design Assurance Level

DS:    *Dossier de Sécurité* (Safety File)

E/E/PE:    *Electrique/Electronique/Programmable Electronique* (Electrics/ Electronics/ Promgrammable Electronics)

EMC:    ElectroMagnetic Compatability

EPSF[16]:    *Etablissement Public de Sécurité Ferroviaire* (Public Rail Safety Establishment)

ESA:    European Space Agency

FAR:    Federal Aviation Regulation

IEC[17]:    International Electrotechnical Commission

ISO[18]:    International Organization for Standardization

JAR:    Joint Aviation Requirement

PSS:    Product assurance and Safety Standards

PLC:    Programmable Logic Controller

RER:    *Réseau Express Régional* (suburban express train)

RAM:    Reliability Avaibility and Maintenability

RAMS:    Reliability Avaibility Maintenability and Safety

RAMSS:    Reliability Availability Maintenability, Safety and Security

SIL:    Safety Integrity Level

SSIL:    Software Safety Integrity Level

STI:    *Spécifications Techniques d'Interopérabilité* (Interoperability Technical Specifications)

STRMTG[19]:    *Service Technique des Remontées Mécaniques et des Transports Guidés* (Technical Service for Installed Mechanics and Guided Transport)

---

16 To learn more, visit http://www.securite-ferroviaire.fr.

17 To learn more, visit: http://www.iec.ch.

18 http://www.iso.org/iso/home.html.

19 To learn more about STRMTG, visit http://www.strmtg.equipement.gouv.fr.

# 4

---

# Risk and Safety Levels

---

From the hazard, we can do some analysis where we analyze the impact on the system. During this analysis, we identify some safety requirements and some safety objectives. These safety attributes can be apportioned on the subsystem, the equipment, the hardware and the software. This chapter describes the general approach of safety management.

## 4.1. Introduction

This chapter aims to explain the notions of risk, feared event, hazard, accident, frequency and hazard severity level. It will set the out basis to understand the differences among the various fields by defining the appropriate vocabulary.

As an example, the aviation sector defines the notion of DAL (design assurance level), whereas other fields use that of SIL (safety integrity level). In the automotive sector, it is the notion of ASIL (automotive SIL) which is used; and two new criteria complement the definition of risk, namely exposure and controllability.

## 4.2. Basic definitions[1]

In the railway domain, the aim is to avoid accidents. An accident (see definition 4.1) is a situation associated with various types of damage (to individuals, to a collectivity).

DEFINITION 4.1 (ACCIDENT – EN 50129).– *an unintended event or series of events that results in death, injury, loss of a system or service, or environmental damage.*

---

1 This section is included in the thesis [BOU 06].

In order to avoid an accident, the possibility of it happening must be considered upon creating the system.

With terrestrial transport systems, as indicated in [BAR 90], accidents may be categorized in different classes:

– *System-initiated accidents*: the users of the system are in a passive position and are a victim of the damage, which can be attributed to failures on the part of the staff (maintenance, intervention, operation, etc.), to failures internal to the system or to anomalous circumstances in the environment;

– *User-initiated accidents*: this category refers to events surrounding one or several users (illness, panic, suicide, etc.);

– *Accidents initiated by a system-user interaction*: as opposed to the previous category, here the user is active and interacts with the system. Incorrect use of the system lies at the root of this type of accident (e.g. non-compliance with the audio signal for door closure).

The notion of an event stands out in our definition of an accident. This notion can by refined by introducing the concept of feared event (see definition 4.2).

DEFINITION 4.2 (FEARED EVENT).– *A feared event is an event which should not occur or can only occur with a very low probability.*

The damage produced can be classified in two distinct families: damage to a collectivity, and damage to an individual.

A user-initiated accident is likely to provoke damage to individuals, whereas the remaining two categories can cause damage either to individuals or collectives. Once the category has been identified, the level of severity of the accident[2] can then be determined: insignificant (minor), marginal (significant), critical and catastrophic.

In [HAD 98], which specifically applies to the railway sector, the authors have created a classification of accidents which highlights the link among the categories, potential accidents, type of damage incurred, severity level, and hazard elements. This classification helps us to make the link between hazard elements such as lightning, and a system-initiated accident, e.g. collision.

As the accident is a final situation, an interesting category is the potential accident. The notion of a potential accident refers to a situation which is known to potentially lead to an accident.

---

2 We have chosen to focus on the railway domain [CEN 99]; however, we have indicated alternative terms between brackets whenever they are used for other sectors.

For rail systems, the list of potential accidents (see definition 4.3) includes: derailment, collision, a passenger falling, getting trapped, electrocution, fire, explosion, flooding, etc.

DEFINITION 4.3 (POTENTIAL ACCIDENT).– *A potential accident is one or a series of unexpected events which may lead to an accident following an additional event which is out of the control of the system.*

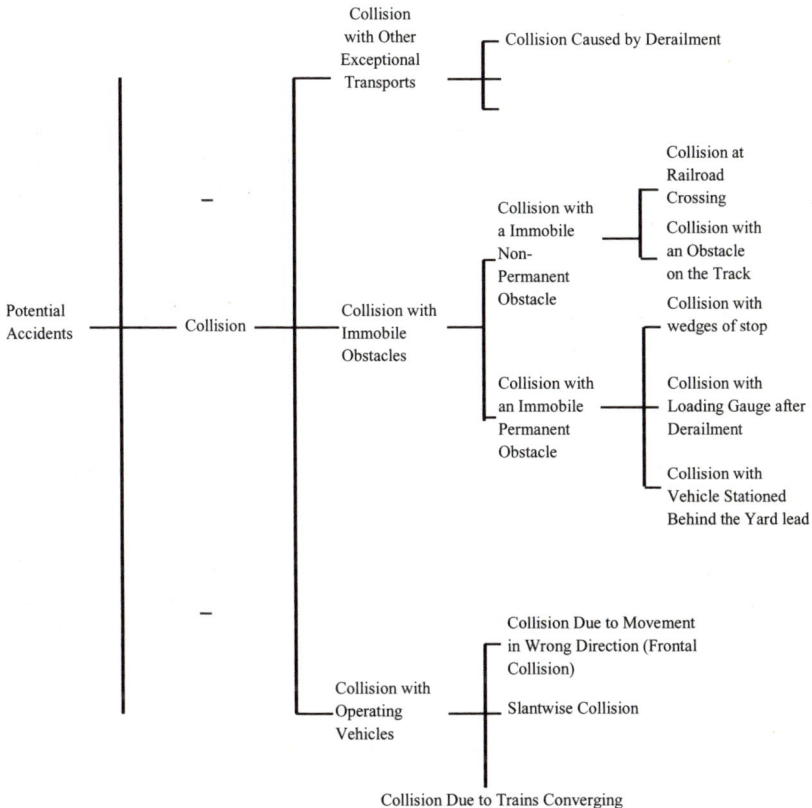

**Figure 4.1.** *Example of a classification of potential accidents*

Figure 4.1 is taken from [HAD 95] and represents an example of the classification of potential accidents with regards to collision. The situation of a potential accident is related to the notion of hazard (see definition 4.4).

DEFINITION 4.4 (HAZARD – EN 50126). – *A physical situation with a potential for human injury.*

For the EN 50129 standard [CEN 99], a hazard is a condition that may result in a potential accident.

A hazard is thus a *dangerous situation* (see definition 4.4), whose consequences can be detrimental to humans (cause injury or death), to society (productivity loss, financial loss, loss of reputation, etc.) or to the environment (degradation of the natural and zoological environment, pollution, etc.). According to whether its origin is random or intentional (deterministic), the terms hazard or threat are used. Threat is used for activities related to safety and/or security.

Hazards can be classified into three families: hazards caused by the system itself or its equipment, hazards provoked by humans (operating staff failure, maintenance staff failure, intervention by the passengers), and hazards resulting from anomalous circumstances in the natural environment (earthquake, wind, fog, heat, humidity, etc.).

Identifying situations which are dangerous for a given system relies on systematic analyses, which implies two complementary phases:

– an empirical phase based on a lessons-learned process (existing lists, accident analyses, etc.);

– a creative and/or predictive phase, which can be based on brainstorming, predictive studies, etc.

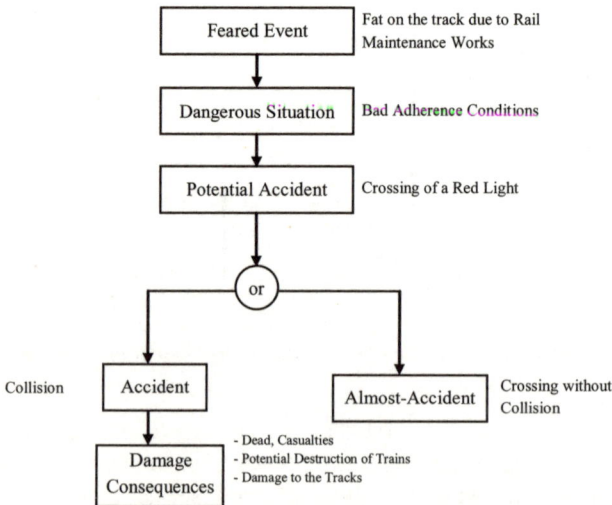

**Figure 4.2.** *Chain of events leading to an accident*

The notion of hazard (dangerous situation) is in direct relation to that of failure.

Figure 4.2 represents the procedure from the feared event to the accident. It is to be noted that the potential accident is only reached in a favorable operational context. Indeed, if a train goes through a red light on a secondary route with little traffic, the risk is lower than that on a main line. Similarly the passage from a feared event to a dangerous situation is related to the technical context.

Figure 4.3 shows a comprehensive diagram of the chain of events, linking them from the cause to the potential accident (to learn more , please refer for instance to [BLA 08]). The notion of cause was introduced in order to link the various types of failures that can affect the system, such as the failure of one or several functions (as a function can be transverse to several parts of internal equipment), failure of one or several pieces of equipment, human error and external factors (EMC[3], etc.).

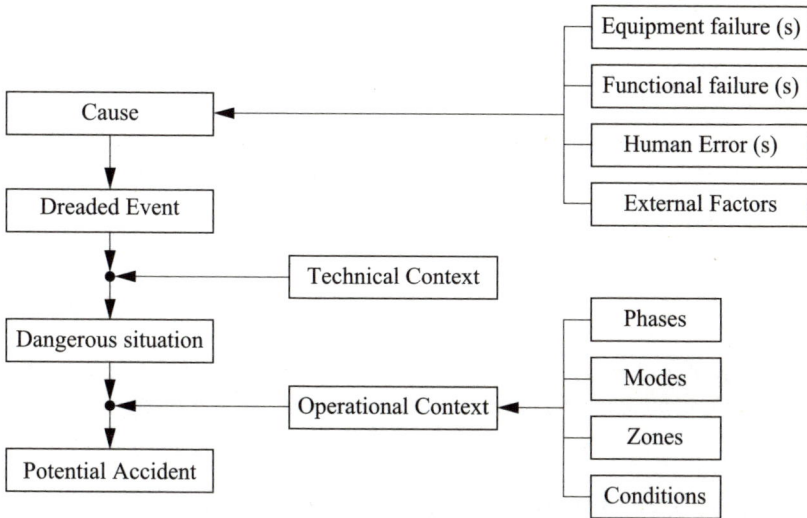

**Figure 4.3.** *Chain of events until the potential accident*

Figure 4.4 shows how, for a potential accident, combinatorial analysis intervenes. It is then the task of analysts to select the representative scenarios (cause/feared event/dangerous situation/potential accident).

---

3 Electromagnetic compatibility (EMC) is the branch of electrical sciences which studies the unintentional generation, propagation and reception of electromagnetic energy with reference to the unwanted effects that such energy may induce.

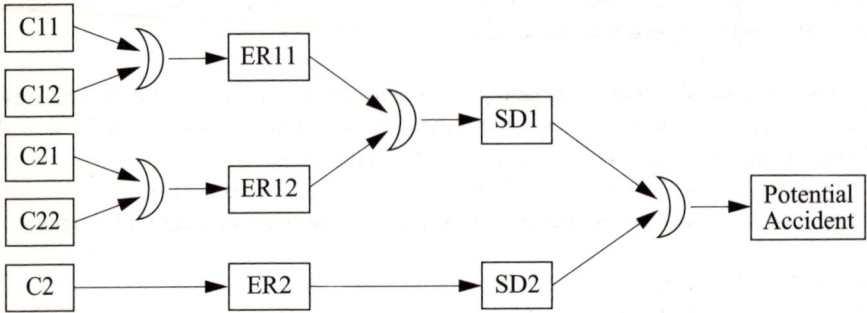

**Figure 4.4.** *Combinatorial analysis of events until the potential accident*

## 4.3. Safety implementation

In this section, we shall discuss and present notions surrounding safety. We shall concentrate mainly on the railway sector, but for some notions, we shall use definitions derived from other fields, if they are relevant.

### 4.3.1. *What is safety?*

In the rail transport sector, two principles of safety implementation are used (see [BIE 98]): probabilistic safety (see definition 4.6) and failsafe safety (see definition 4.5).

DEFINITION 4.5 (FAILSAFE SAFETY).– *A system is said to be intrinsically safe if it has been ascertained that any failure of one or several components cannot push it to a more permissive state than that in which it is when the failure occurs. It is to be noted that for rail transport, a complete halt is generally the most restrictive state.*

Failsafe safety is based on physical properties of the components, so that it is necessary to know the behavior of these components in the event of one or several errors. This is why only the components whose complexity can be managed are used.

DEFINITION 4.6 (PROBABILISTIC SAFETY).– *Probabilistic safety consists of demonstrating a priori that the probability for a dangerous situation to arise is lower than a predefined threshold.*

The concept of probabilistic safety admits that there is no such thing as zero risk. This concept will be fully described later. The combination of probabilistic and failsafe safety leads us to the definition of "tested safety". A device is said to be in tested-safety mode if it is able to detect errors leading to a state of seriousness with unacceptable consequences, and if, in that event, it commands a safe configuration.

DEFINITION 4.7 (COMPUTER SECURITY).– *Computer-based security is defined as all the methods deployed to minimize vulnerability to accidental or intentional threats in a computer system.*

Two different types of threats can arise, namely accidental and malicious threats. Two different terms are used in English, safety (innocuousness) and security (privacy).

Safety refers to the protection of computer systems against accidents caused by the environment or flaws in the system. It is relevant in the context of computer systems controlling real-time processes and putting human lives on the line (computer-based transport, nuclear power plants, etc.). Real-time systems are under strict time constraints.

DEFINITION 4.8 (SAFETY).– *Safety refers to disaster prevention (innocuousness).*

In standards EN 50126 and EN 50129 [CEN 99, CEN 03], safety is defined as freedom from unacceptable risk of harm.

Protecting computer systems against malicious actions (intrusions, vandalism, etc.) leads to security. Generally speaking we are facing systems processing sensitive data (bank account management, etc.).

Since the introduction of new technologies within computer-based systems, this is no longer entirely true. With rail transport, WiFi or GSM-based technologies (GSM-R[4] or otherwise) are being introduced and security measures will have to be taken.

DEFINITION 4.9 (SECURITY).– *Security refers to the prevention of attacks (privacy).*

---

4 GSM-R stands for Global System for Mobile communications – Railways; see http://www.gsm-rail.com.

### 4.3.2. *Safety management*

The dependability of a system is achieved through a number of activities based on studies. These studies must be performed before designing the system in question.

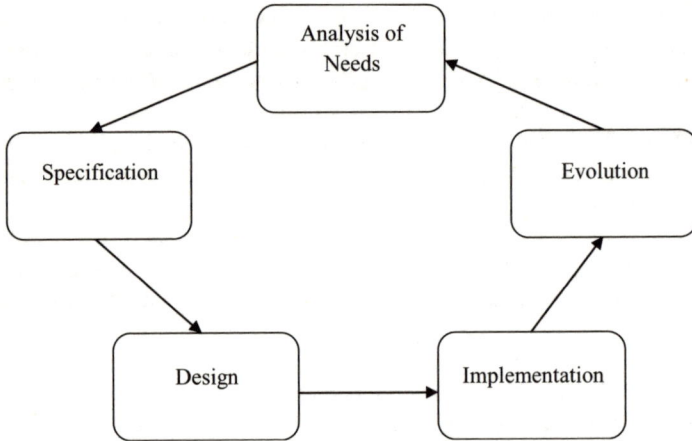

**Figure 4.5.** *Implementation cycle of dependability*

Figure 4.5 shows the implementation of dependability for a computer-based system. The first phase consists of rigorously analyzing the needs. Studies related to dependability should help in understanding and identifying the risks, and thinking of possible consequences. From this, dependability objectives can be defined (reliability, availability, maintenance and safety) for our system.

Studies related to dependability follow the different phases of the system's lifecycle (specification, design, development, and evolution).

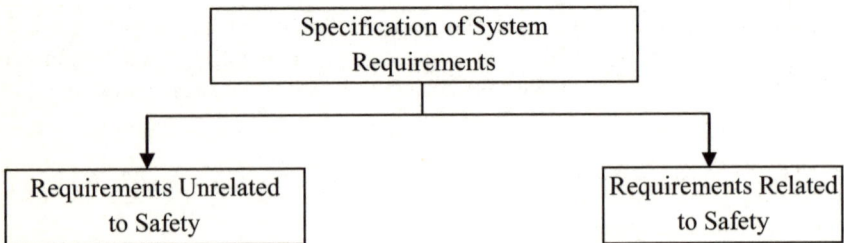

**Figure 4.6.** *Specification of the system requirements*

The specification of requirements is an important step as it helps identifying the needs and showing the requirements of the system. These requirements are of two different types: those unrelated to safety (functional requirements, non-functional requirements such as performance, etc.) and the requirements related to safety (see Figure 4.6).

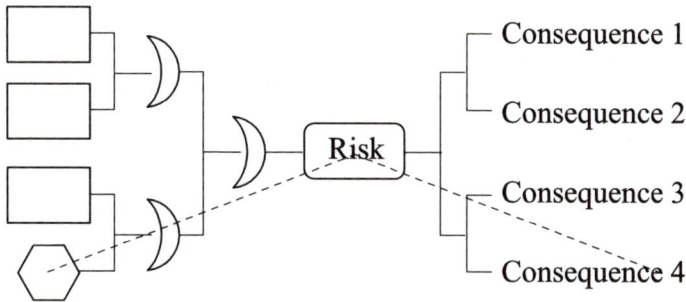

**Figure 4.7.** *Relationships among failures, risks and effects*

As shown in Figure 4.7, it is possible to establish a link between a risk and its consequences and the failures associated with it.

DEFINITION 4.10 (RISK).– *A risk is a combination of:*

*– The probability of occurrence of one or a combination of events leading to a dangerous situation, or the frequency of such events.*

*– The consequences of this dangerous situation.*

In the light of definition 4.10, and in compliance with [IEC 08], the risk is a combination of the probability and seriousness of the damage. Severity is characterized by the damage and consequences caused (see Figure 4.8).

For each dangerous situation, the probability of occurrence or frequency must be determined. In order to do this, a set of categories can be defined (see Table 4.1 taken from the CENELEC EN 50126 standard [CEN 99]), and each category is associated with a frequency range.

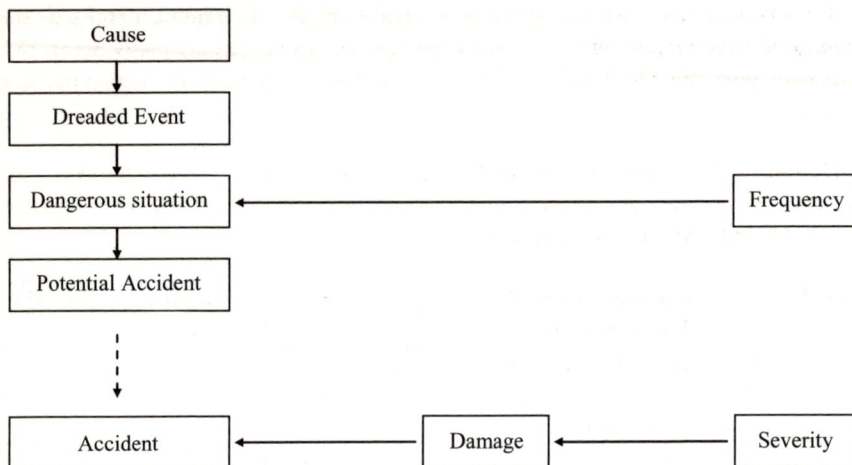

**Figure 4.8.** *Frequency and seriousness*

In Table 4.1, the term often indicates that the dangerous situation has occurred and will arise again.

| Category | Description |
|----------|-------------|
| Frequent | Likely to occur frequently.<br>The situation is dangerous and present continuously. |
| Probable | Will occur repeatedly.<br>It can be expected that the dangerous situation occurs often. |
| Occasional | Likely to occur several times.<br>The dangerous situation can be expected to arise. |
| Remote | Likely to occur sometime in the system lifecycle.<br>The dangerous situation can reasonably be expected to occur. |
| Improbable | Unlikely to occur but possible.<br>The dangerous situation can be assumed to occur in exceptional circumstances. |
| Incredible | Extremely unlikely to occur.<br>It can safely be assumed that the dangerous situation will not occur. |

**Table 4.1.** *Probability of occurrence or frequency of dangerous situations*

DEFINITION 4.11 (DANGEROUS SITUATION – IEC 61508).– *A dangerous situation is a situation in which a person is exposed to one or several dangerous phenomena. This dangerous phenomenon is a potential source of hazard in the short and/or the long term.*

Definition 4.11 introduces a link between the risk and a potential accident. The dangerous situation is one of the elements of the accident scenario. It corresponds to an unstable but reversible state of the system.

| Level | Consequences for People or the Environment | Consequences for the Service |
|---|---|---|
| Catastrophic | Fatalities and/or multiple severe injuries and/or major damage to the environment | |
| Critical | Single fatality and/or severe injury and/or significant damage to the environment | Loss of a major system |
| Marginal | Minor injury and/or significant threat to the environment | Severe System(s) damage |
| Insignifiant | Possible minor injury | Minor damage to the system |

**Table 4.2.** *Level of seriousness in dangerous situations*

For each dangerous situation, the consequences for the system, the people and the environment should be analyzed. Quantification can be achieved through defining the level of seriousness. Table 4.2 is derived from the CENELEC EN 50126 standard and links dangerous situations to their consequences.

Given the diversity of the risks and the impacts these can have on the system, it is preferable to define categories and associate actions with them. Table 4.3 is an example of such a categorization. For intolerable categories, there is a link with the supervisory authority. In the railway sector, this authority is the operating company or the railway authority (in each country there is at least one railway authority).

| Risk Category | Actions to be Associated to Each Category |
|---|---|
| Intolerable | Will be eliminated |
| Undesirable | Will only be accepted when risk reduction is impracticable and with the agreement of the railway authority or the safety regulatory authority, as appropriate |

**Table 4.3.** *Qualitative categories of risk*

| Tolerable | Acceptable with adequate control and with the agreement of the railway authority |
| Negligible | Acceptable with/without agreement of the railway authority |

**Table 4.3 (continued).** *Qualitative categories of risk*

In order to assess the level of risk following the definitions, standards use an occurrence-seriousness matrix. As an example, we have taken the matrix (Table 4.4) which is used in the standard CENELEC EN 50126.

| Frequency of occurrence of a dangerous event | Level of risk (risk categories) | | | |
|---|---|---|---|---|
| Frequent | Undesirable | Intolerable | Intolerable | Intolerable |
| Probable | Tolerable | Undesirable | Intolerable | Intolerable |
| Occasional | Tolerable | Undesirable | Undesirable | Intolerable |
| Remote | Negligible | Tolerable | Undesirable | Undesirable |
| Improbable | Negligible | Negligible | Tolerable | Tolerable |
| Incredible | Negligible | Negligible | Negligible | Negligible |
| | Insignifiant | Marginal | Critical | Catastrophic |
| | Severity Levels of Hazard Consequence | | | |

**Table 4.4.** *Qualitative risk categories*

NOTE.– The content of Tables 4.1, 4.2 and 4.3 is directly taken from standards. The wording of the elements depends on the general character of their application. With each new implementation, the terms and categories are defined first.

Acceptable risk is a value associated with risk resulting in explicit objective decisions. The acceptability threshold is referred to as the tolerable hazard rate (THR). The THR is the probability of occurrence of a failure expressed as $10^{-x}$ per hour. When identifying dangerous situations, a THR must be associated with it and will be part of the specification.

For a particular system, the responsibility of providing the maximum acceptable rate of occurrence of hazard (THR) lies with the operating company. The safety management process for the railway sector is defined in the CENELEC EN 50126 standard [CEN 99] and is presented in Figure 4.5.

Managing the risks with a capacity going beyond the acceptability threshold involves:

– decreasing their probability of occurrence through preventive measures that reduce the vulnerability in the most exposed elements of the system;

– decreasing the severity of the consequences through protective measures.

From an initial description of the system and a list of dangerous situations, it is possible to carry out a risk analysis that identifies the measures to be taken to allow only acceptable risks. These measures must be included in the conception of the system.

Figures 4.9 and 4.10 show an example of a risk analysis process. This process can be applied to the system, the subsystems and equipment. The identification phase for dangerous situations consists of taking the interface of the system and identifying the dangerous situations as well as the THR associated with each of them.

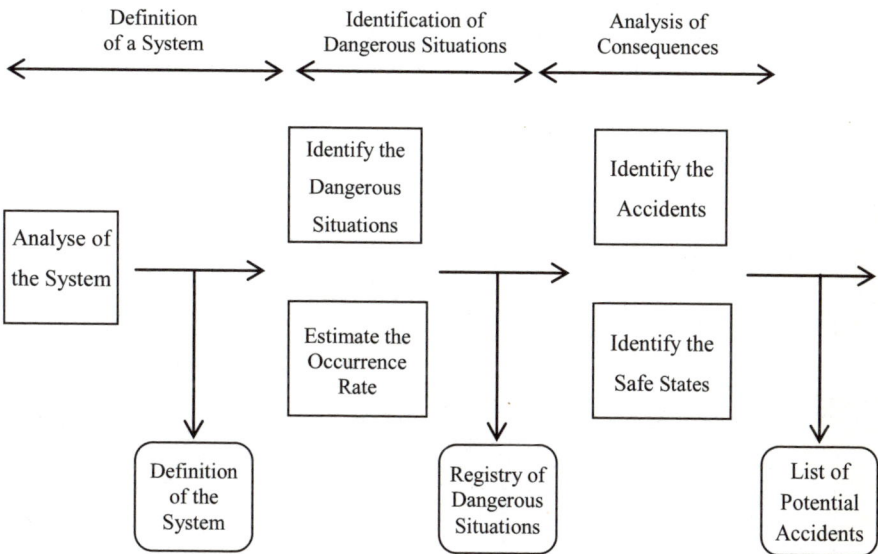

Definition of a System → Identification of Dangerous Situations → Analysis of Consequences

Analyse of the System →

Identify the Dangerous Situations

Estimate the Occurrence Rate

Definition of the System

Identify the Accidents

Identify the Safe States

Registry of Dangerous Situations

Identify the Accidents

List of Potential Accidents

**Figure 4.9.** *Example of risk analysis process – part one*

When it comes to risk, if hazards can be eliminated, they must be, as this is the most effective way of decreasing risk. However, eliminating hazards completely is often unfeasible. In these cases, risks must be reduced to acceptable levels.

Risk Estimation            Allocation         Management of
                                              Dangerous Situations

Determine                  Compare with
the Risk                   Objectives

List of          List of        Specification
Potential        Risks          of Safety          THR
Accidents                       Requierements

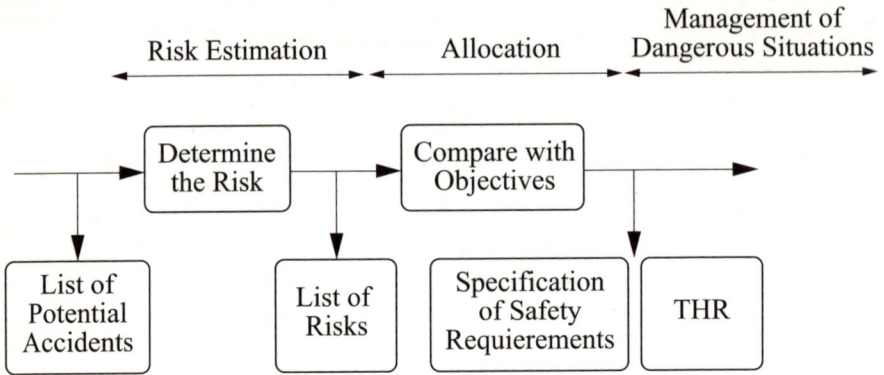

**Figure 4.10.** *Example of risk analysis process – part two*

In order to lower risks, the frequency of a hazard must be reduced and/or the seriousness of the hazard (consequences). Decreasing the frequency and/or the seriousness of the hazard is more commonly referred to as "risk management".

Figure 4.11 is a diagram of the risk management process as an iterative loop divided into three phases: identification, evaluation and action.

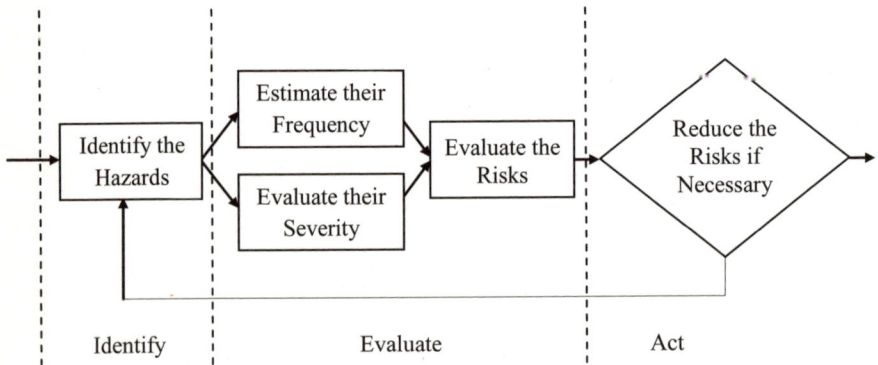

Identify the      Estimate their      Evaluate the      Reduce the
Hazards           Frequency           Risks             Risks if
                                                        Necessary
                  Evaluate their
                  Severity

Identify              Evaluate                Act

**Figure 4.11.** *Risk management process*

Risk management:

– stage 1: identifying the hazards associated with the system:

- identifying and organizing in a systematic hierarchy all the reasonably predictable hazards associated with the system in its environment,

  - identifying the sequences of events leading to hazards;

– stage 2: assessing the hazards:

  - estimating the frequency of occurrence of each hazard,

  - assessing the probable severity of the consequences of each hazard,

  - evaluating the risk of each hazard for the system;

– stage 3: compiling a process for the management of the risk underway:

  - compiling a register of dangerous situations[5] (RDS).

Once a list of safety requirements of THR is compiled, it is then possible to define the level of safety integrity.

### 4.3.3. *Safety integrity*

In other fields (aviation, nuclear, aerospace, etc.), a similar notion is used, i.e. the notion of a safety level. The notion of safety integrity was introduced for the first time in the IEC 61508 standard ([IEC 08]).

The notion of safety integrity was then taken up and divided into different daughter standards. For the railway sector, the notion of safety level is found in the standards CENELEC EN 50126, EN 50128 and EN 50129.

The notion of safety integrity (see Figure 4.12) is made up of two components:

– integrity in the case of random failures,

– integrity in the case of systematic failures.

The main difference between random and systematic failures lies in the fact that probabilities cannot be calculated for the latter.

Random failures originate in the hardware, appear at random intervals, and are due to the ageing and wearing of hardware elements. Software applications on the other hand are not subject to random failures.

---

5 We also refer to a safety diary and/or a HazardLog (HL).

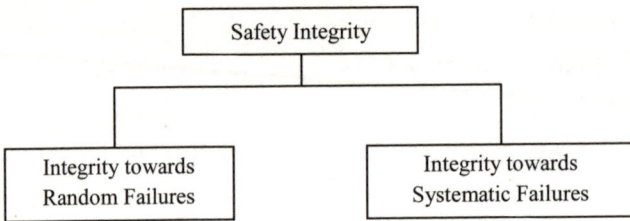

**Figure 4.12.** *Safety integrity*

Systematic failures can originate in the hardware and/or software. They are caused by human errors during the various phases of the system's lifecycle (specification and/or design problems, manufacturing problems, problems during the system's implementation, or introduced during maintenance work, etc.). Systematic failures require different methods or techniques to reach a satisfactory level of confidence within a particular level of safety integrity.

Figure 4.6 shows a detail of the requirement specification in two families; the requirements related to safety and those unrelated to safety. In light of the above, it is possible to refine this branching as seen in Figure 4.13.

This figure introduces a new way of partitioning safety-related requirements between functional safety and safety integrity requirements.

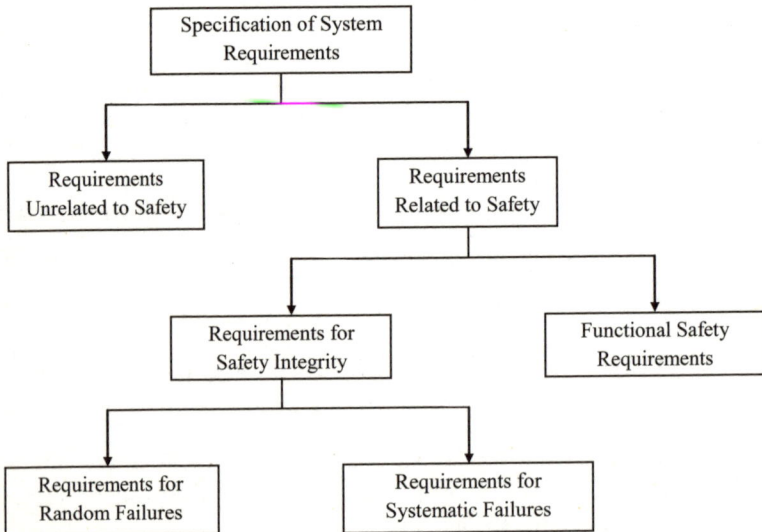

**Figure 4.13.** *Specification of the system requirements*

Functional safety requirements characterize needs related to the operational side of the safety-related functions in the system.

The notion of safety integrity, on the other hand, can be characterized as the capacity of a system to fulfill its safety functions in specific conditions, within a defined operational environment within a limited period of time. Safety integrity (section 4.7 of standard CENELEC EN 50126 [CEN 99]) refers to the probability of not fulfilling a required safety function.

The CENELEC EN 50126 standard (which applies to the whole rail system), as opposed to the IEC 61508 standard ([IEC 08]), does not define the correlation between safety integrity and failure probabilities. However, these must be defined for a specific application. Safety functions are assigned to safe systems or external devices for risk reduction.

The safety integrity level (SIL) can be divided into four discrete values (from 1 to 4) which help specifying prescriptions for safety integrity.

In the IEC 61508 standard ([IEC 98]), the four levels of SIL for a given safety function are characterized by the impact of failures:

– SIL 4: catastrophic impact (highest level);

– SIL 3: impact on the community;

– SIL 2: major protection of implant and production needed, or else risk of injury for the staff;

– SIL 1: minor protection of implant and production needed (lowest level).

The SIL must be determined by experts and assigned to an element that fulfills one or several isolated functions.

### 4.3.4. *Determining the SIL*

A systematic approach is necessary (see Figures 4.9 and 4.10) to determine the safety requirements. This approach should include the environment, the various operating modes and the architecture of the element under scrutiny.

Once the relevant risks and dangerous situations have been identified, this result in a list of hazards with the THR associated with them.

Appendix A of the standard EN 50129 ([CEN 03]) introduces the methods for dividing the THR and assigning its daughters to each of the feared events:

– Step 1: analysis of the causes of each feared event so as to identify the system's functions whose failure leads to the feared failures.

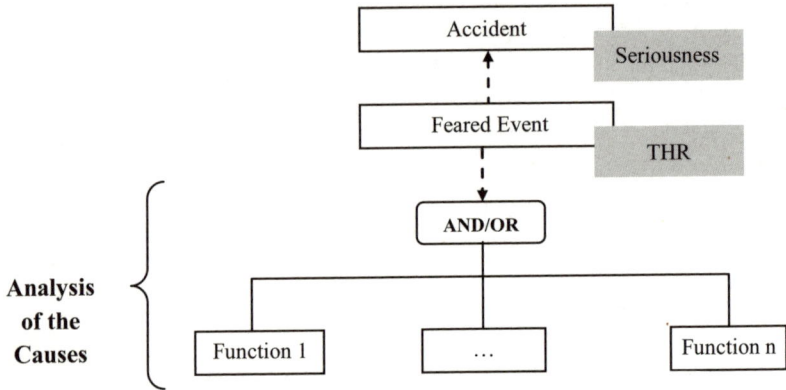

**Figure 4.14.** *Analysis of the causes associated with each feared event*

– Step 2: division of the THR associated with the feared event and of the THR associated with the system's functions according to the combinatorics operators AND/OR.

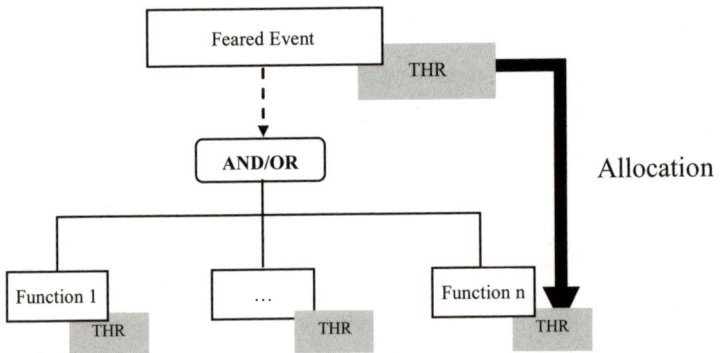

**Figure 4.15.** *Allocation of THR according to a failure trees*

– Step 3: allocation of SIL to the system functions using the corresponding THR (see Table 4.6).

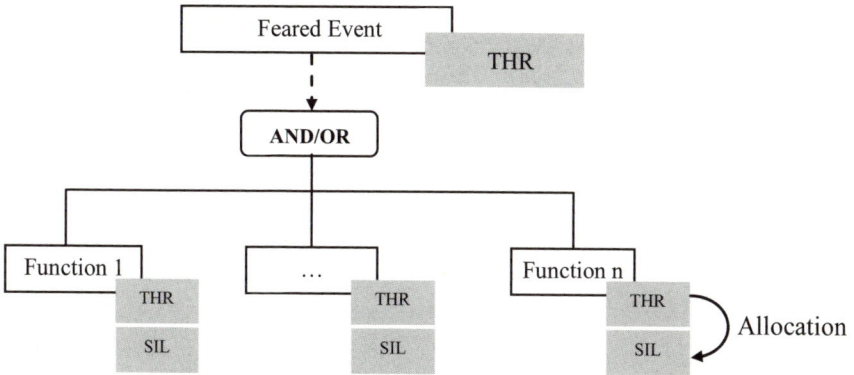

**Figure 4.16.** *Allocation of SIL on the basis of THR*

– Step 4: division of the THR and SIL level for each of the system's functions to fit the different subsystems supporting the function.

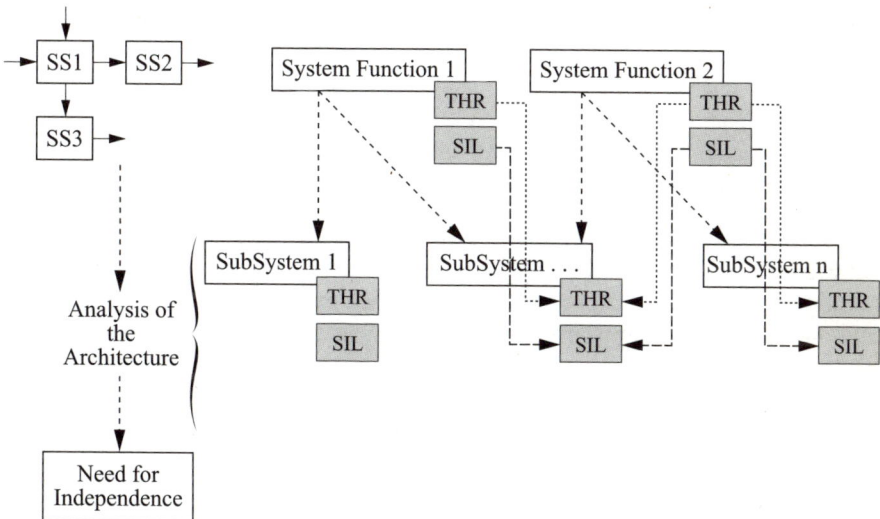

**Figure 4.17.** *Allocation for subsystems*

Figure 4.17 shows, on the left side, the analysis of the architecture which has architecture for inputs, a list of the system functions with their THR, as well as the link between the system and the subsystem functions.

The right side of Figure 4.17 highlights the link between the functions of the systems and those of the subsystems. If a given subsystem function is used only by a system function, then there is a direct allocation of the THR to SIL, which is the same at the system and subsystem levels. If a subsystem function supports several system functions, then the SIL of that system function equals the maximum SIL for the subsystem functions; and the THR for that system function equals the minimum THR for the subsystem functions.

Analysis of the causes of failures should support this allocation, as the THR for the subsystem functions are calculated through the tree. When it comes to the SIL allocation:

– For functions whose safety-hampering failures are combined through an "OR" junction in a causal tree are allocated the same SIL level than that which was determined upstream of the "OR" branching.

– Allocation of SIL to the functions whose safety-hampering failures are combined through an "AND" junction in the causal tree relies on respecting the functional and physical independence criteria for those functions. If the functional and physical independence of these functions is clearly established, the SIL which was determined upstream of the "AND" junction can be divided among the functions on each side of this "AND" junction. In the opposite case, the upstream SIL should be allocated to the functions.

– Step 5: division of the THR allocated to each subsystem in FR (failure rate), to hardware equipment making up the subsystem according to its architecture (assumptions of independence can then be made); and division at the level of SIL allocated to each subsystem in SIL (SSIL) assigned to hardware equipment (software):

- the level of hardware SIL is identical to that of the subsystem SIL;

- the level of SSIL (software SIL) for the software is identical to the level of SIL for the subsystem.

It is to be noted that we use the hypothesis that a subsystem is composed of a "hardware" and a "software" part.

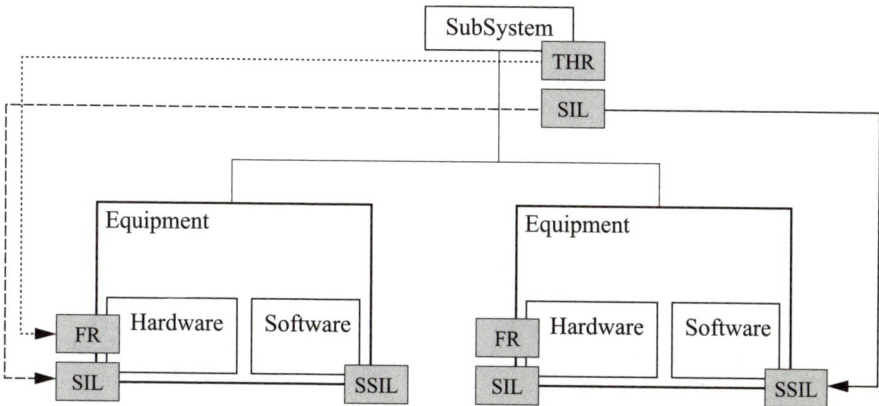

**Figure 4.18.** *Division of SIL and THR for the subsystem between pieces of equipment*

The passage from a subsystem SIL to an equipment SIL is regulated, however it is possible to refine these SIL and SSIL by taking into account the architecture and equipment.

If we use a homogeneous 2oo3 architecture (same hardware elements, same OS[6], same memory type, etc.), random flaws are better included, but not the design or common flaws (processor, etc.). This is why the required level of SIL for the hardware should be the same as that for the subsystem.

If on a given architecture, the same software is used three times, this must be SSIL4. If it can be diversified and we are able to show diversity in software applications, the use of three different software packages can justify a SSIL2 on those packages.

The choice of a heterogeneous 2oo3 architecture (three different processors, different memory cells, etc.) helps improve the management of design problems and flaws in the basic components, while limiting common modes. It is then possible to show that each treatment unit can be attributed to a SIL2 objective. If we use a single software application, it will have to remain SSIL4.

---

6 OS stands for operating system.

The arguments used as examples show that it is possible to refine the SIL and SSIL by taking into account the architecture and an expert judgment. It is to be noted that the allocation of SIL and SSIL requires expertise and must be appropriately justified. It is by no means an algorithm of SIL.

### 4.3.5. SIL table

The IEC 61508 standard ([IEC 08]) defines two categories of systems: high demand or continuous mode of operation systems and low demand mode of operation systems. In order to characterize these two types of systems, the standard identifies the notion of mean probability of failure on demand (PFDavg) as well as the notion of probability of failure per hour (PFH). Table 4.5 presents the definitions of SIL for both types of systems.

| Low demand mode of operation | | High demand or continuous mode of operation | |
|---|---|---|---|
| Safety Level | Mean Probability of Failure on Function Demand | Safety Level | Probability of Dangerous Failure per Hour |
| SIL 4 | $10^{-5} \leq PFDavg < 10^{-4}$ | SIL 4 | $10^{-9/h} \leq PFH < 10^{-8/h}$ |
| SIL 3 | $10^{-4} \leq PFDavg < 10^{-3}$ | SIL 3 | $10^{-8/h} \leq PFH < 10^{-7/h}$ |
| SIL 2 | $10^{-3} \leq PFDavg < 10^{-2}$ | SIL 2 | $10^{-7/h} \leq PFH < 10^{-6/h}$ |
| SIL 1 | $10^{-2} \leq PFDavg < 10^{-1}$ | SIL 1 | $10^{-6/h} \leq PFH < 10^{-5/h}$ |

**Table 4.5.** *Table of SIL from the IEC 61508 standard [IEC 08]*

For the railway sector, Table 4.6 shows the existing link between SIL and THR. This table was actually introduced for rail signaling and is extendable to part or the entire system. This table can also be redefined.

| THR per Hour and per Function | SIL |
|---|---|
| $10^{-9} \leq 9THR \leq 10^{-8}$ | 4 |
| $10^{-8} \leq 8THR < 10^{-7}$ | 3 |
| $10^{-7} \leq 7THR < 10^{-6}$ | 2 |
| $10^{-6} \leq 6THR < 10^{-5}$ | 1 |

**Table 4.6.** *Table of SIL from the CENELEC EN 50129 standard [CEN 03]*

NOTE.– It is to be noted that Tables 4.5 and 4.6 should be read from left to right. A given THR then allows a SIL level to be defined. The SIL level corresponds to the level of confidence to be achieved when correcting the system. In some exceptional cases, it is possible to go from a SIL level to a THR level. In those few cases, the level of safety for a hardware element is introduced in order to attain the SIL objective. As this hardware element must be quantified, a THR is necessary.

### 4.3.6. *Allocating SIL*

In the previous sections, we have shown the general methodology for allocating SIL and SSIL. However, as indicated in the explanations for step 5 in section 4.3.4, the choice of SIL and SSIL should be made with caution and by an expert, and implies taking into account identified risks, safety analyses (preliminary analyses of risks, FMEA, failure trees, software error effect analysis[7], etc.), objectives such as THR and the proposed architecture.

| Global SIL | SIL Low-Level Functions | | Combination Function[8] |
|---|---|---|---|
| | Main | Other | (e.g. voter) |
| SIL4 | SIL 4 | None | None |
| | SIL 4 | SIL 2 | SIL 4 |
| | SIL 3 | SIL 3 | SIL 4 |
| SIL 3 | SIL 3 | None | None |
| | SIL 3 | SIL 1 | SIL 3 |
| | SIL 2 | SIL 2 | SIL 3 |
| SIL 2 | SIL 2 | None | None |
| | SIL 1 | SIL 1 | SIL 2 |
| SIL 1 | SIL 1 | None | None |

**Table 4.7.** *SIL allocation table from the Yellow Book [RSS 07]*

There are several approaches to defining the general rules for architecture, such as the Yellow book[9] ([RSS 07]).

Other approaches, such as [HAR 06, PAP 10], take it one step further as they suggest that allocating SIL could be done automatically.

---

7 SEEA (software error effects analysis) is the only study to be carried out over the whole or part of a given software application.

8 In general, the combination function (e.g. voter) has the same safety level as the global function, but is less complex.

9 The Yellow Book describes how the standards CENELEC EN 50126, EN 50129 and EN 50128 should be implemented in the UK.

### 4.3.7. *SIL management*

In the same line as for the standard IEC 61508 ([IEC 08]), standards CENELEC EN 50126, EN 50128 and EN 50129 detail the requirements to be met for each level of safety integrity. These requirements are more rigorous for the highest levels of safety integrity so as to ensure a lower probability of failure.

As shown in Figure 4.19, safety integrity is managed through two groups of practices. When integrity addresses random failures, quantitative (failure rate) and qualitative objectives are managed. As for systematic failures, objectives of quality management, safety management and safety management of the hardware elements are included.

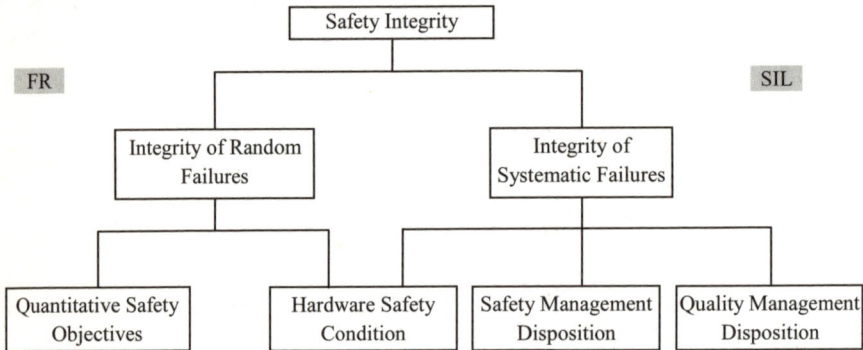

**Figure 4.19.** *Safety Integrity Management*

We shall now focus on managing integrity for systematic failures:

– measures taken for the management of hardware elements (see IEC 61508 part 2, CENELEC EN 50129) are based on safety analysis methods (PHA, FMEA, HAZOP, FTA, etc.), design methods[10] (redundancy, diversification, etc.), management of degraded mode operations, definition strategies and management of technical documents;

– as for quality management (see IEC 61508 part 1, CENELEC EN 50129 and CENELEC EN 50128), the measures rely on quality control (exemplified in ISO 9001:2008 or similar), control of staff competence and implementation of independence;

---

10 For more information, please consult Chapter 1 in [BOU 09].

– measures for safety management (see IEC 61508 parts 1 and 3, CENELEC EN 50126, EN 50129 and EN 50128) find their ground in safety control as a whole (definition of a strategy which covers from the hardware through to the hardware elements and software), safety demonstration, management of software application safety (use of SEEA, of controlled development methods and of "safe" programming techniques) as well as control of the technical documentation associated to each of the different activities (an audit should be easily compiled for each activity).

### 4.3.8. *Software SIL*

As mentioned in section 4.3.3, there are four SIL (safety integrity levels). When an element is not subject to SIL, no level can be attributed to it, so that the standard cannot be applied for that element.

On the other hand, for software, SSIL is used (software safety integrity level) with its five different values, from 0 to 4. Level 4 corresponds to the highest safety integrity level, whereas level 1 is the lowest one. Level 0 is used to indicate that the software in question does not interfere with safety issues.

The main difference between SIL and SSIL is that, for SSIL 0, the standard CENELEC EN 50128 [CEN 01a, CEN 11] identifies a number of obligations.

It is to be noted that CENELEC EN 50128 [CEN 01a, CEN 11] is formulated in such a way that SSIL can be envisaged as having only three levels:

– SSIL 0: the software does not have safety implications but its quality must be under control and its configuration managed;

– SSIL 2: the software is associated with a "medium" safety objective which requires the application of principles ensuring safety;

– SSIL 3-4: the software is associated with a "high" safety objective which requires the deployment of appropriate measures and techniques.

The CENELEC EN 50128 standard [CEN 01a] identifies tools and methods for a "flawless" software. Below is an overview of the techniques recommended in CENELEC EN 50128:

– using formal methods;

– using dynamic tests;

– using development workshops to develop qualifications;

– using simulation methods to validate the model and/or to select the tests;

– formalizing the working methods of the various teams (specification, coding, verification, validation, etc.);

– introducing a high quality assurance (systematically pre-established quality for the system).

### 4.3.9. *Iterative process*

The process of the SIL definition that starts from the hazards and proceeds with the lowest-level elements (software and hardware) is a process that requires loop connection, as shown in Figure 4.20. The purpose of this loop is to verify that all the predefined objectives are met.

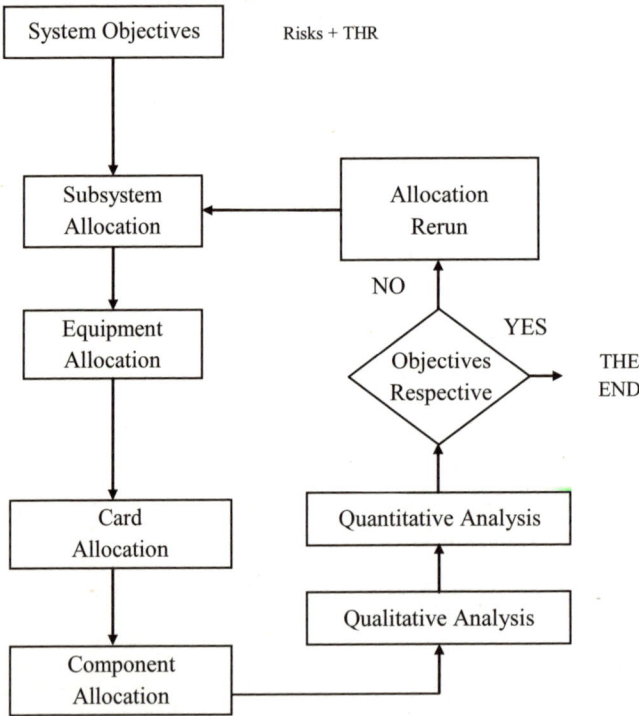

**Figure 4.20.** *Allocations of the system to the basic elements*

### 4.3.10. *Identifying safety requirements*

Figure 4.13 introduces safety-related requirements; however there is no mention of how the link between risk analysis and safety requirements was established. Risk management identifies the barriers that protect from a potential accident and/or help in reducing the seriousness of the accident's consequences.

Figure 4.21 shows the barriers within a sequence of events leading to an accident. Generally speaking, the barriers are situated in the equipment, the system functions and/or in particular operating practices.

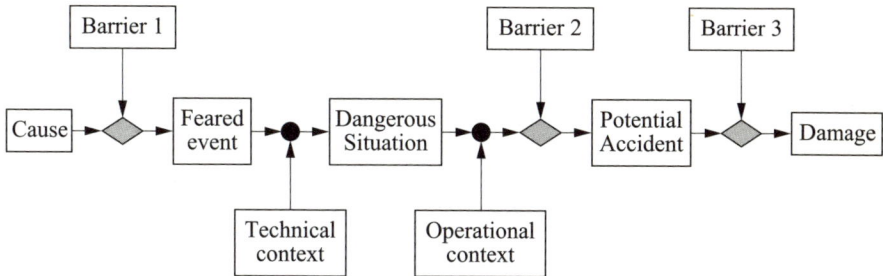

**Figure 4.21.** *Introduction of barriers*

Safety analyses should define the safety requirements that will be backed by the barriers. Figure 4.22 reveals a number of elements (gray boxes) which can support safety requirements.

It is to be noted that the process starts with risk identification, then goes on to determine the safety-related functions of the system. It divides these system functions to fit the various pieces of equipment so as to identify the equipment that supports safety-related functions. Eventually, each piece of equipment is associated with a component of requirements, which is a function of the risk incurred.

It is to be noted that one of the main points for this type of analysis is the delivery of choices and the justifications that go with them.

The SIL can thus be defined as the level of confidence granted to the respect of safety requirements allocated to the function and/or equipment.

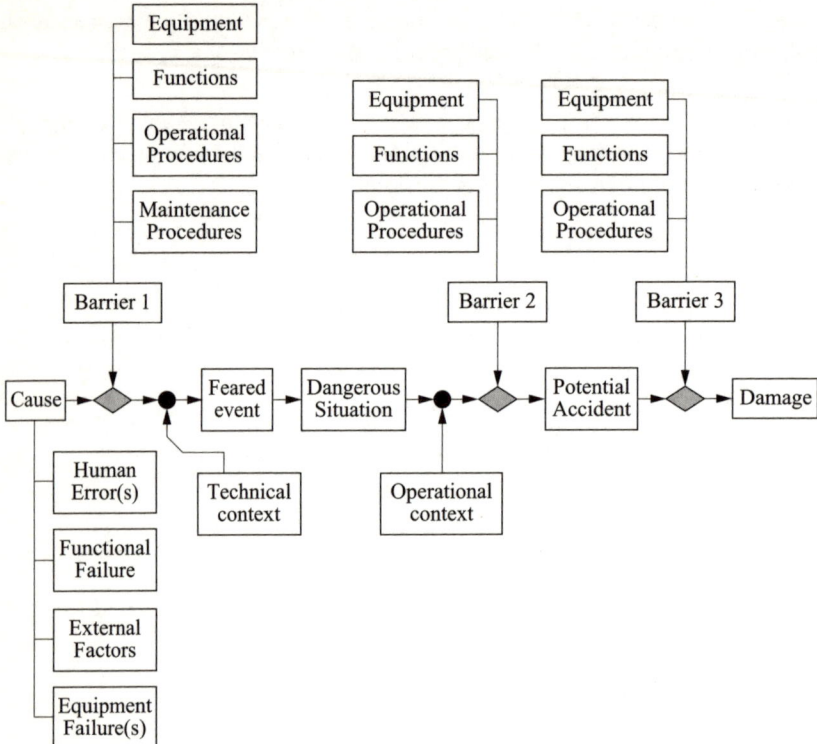

**Figure 4.22.** *Positioning of safety requirements*

## 4.4. In standards IEC 61508 and IEC 61511

[ISA 05] gives a possible approach and a few rules for applying standards IEC 61508 [IEC 98] and IEC 61511 [IEC 05].

Part 5 of IEC 61508 [IEC 98] describes the risk diagram as a means of identifying the SIL. The IEC 61511 standard [IEC 05] in part 3 gives two methods to calculate the SIL: the risk diagram and the LOPA (layer of protection analysis). Readers can learn more about these methods by referring to [GUL 04] and [AHM 09].

These three methods of SIL determination for a given instrumental function are to be applied after a risk analysis. It is to be noted that contrary to Appendix A of CENELEC EN 50129 [CEN 03], which is *normative*, Appendix D, E and F in part 3 of IEC 61511 (same for standard IEC 61508) are *informative*, i.e. there is no obligation to follow them.

### 4.4.1. *Risk diagram*

One of the most widely used qualitative methods of SIL determination for instrumental system safety is the "risk diagram" method (see standards IEC 61508 [IEC 98] and IEC 61511 [IEC 05]).

Adopting this method requires the introduction of a number of simplified parameters to describe the nature of the dangerous situation in the event of safety-related system failures or unavailability.

As previously indicated, a risk R is perceived as a combination of a frequency f and seriousness, which we shall refer to as consequences C. The frequency of the dangerous event f is supposed to result from three influencing factors:

– frequency and duration of exposure in a dangerous zone;

– the possibility of avoiding a dangerous event;

– the probability that the dangerous event takes place without any safety-related system, sometimes called the probability of undesired occurrence.

If we combine the fact that $R = f * C$ and that $f = F * P * W$, then we can characterize risk through four parameters:

– consequences of the dangerous event (C);

– frequency and duration of exposure to the hazard (F);

– possibility that the dangerous event will be avoided (P);

– probability of undesired occurrence (W).

For a particular hazard, each parameter should be analyzed and associated so as to decide the level of SIL for safety-related systems.

These four parameters are generic enough to be applied to all applications and they help organize risks into a significance rating. It is possible to introduce new parameters in order to refine this methodology.

Figure 4.23 shows an example of a risk diagram. In the scales associated with parameter W, the diagram reveals six classes of requirement, rated from "a" to "b" by going through SIL 1 to SIL 4. Category "a" corresponds to "no safety requirements", whereas category "b" relates to an "unacceptable situation".

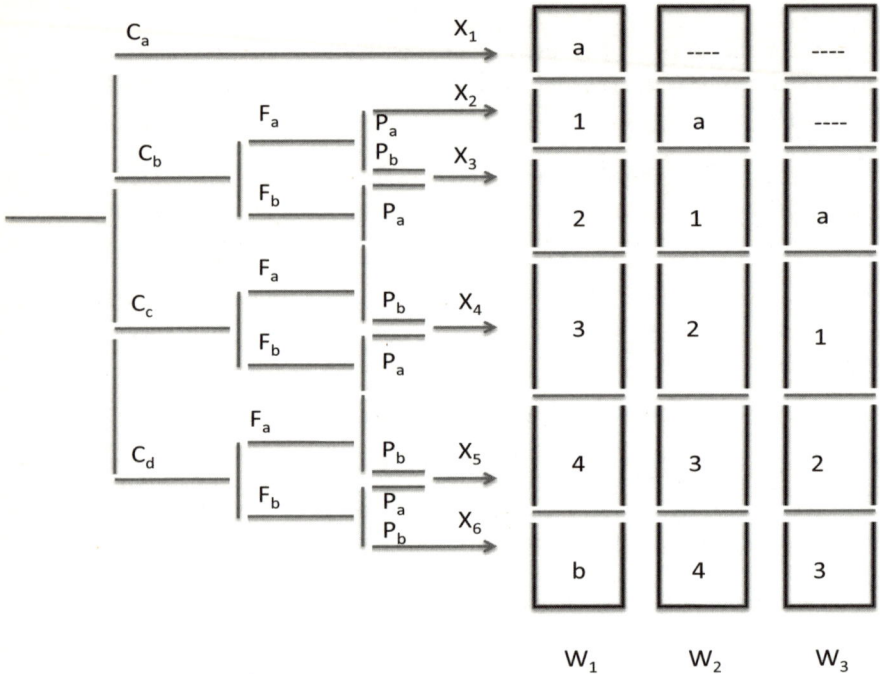

**Figure 4.23.** *Example of a risk diagram*

For a specific output in a risk diagram (i.e. X1, X2, ... or X6) and for a specific W scale (i.e. W1, W2 and W3), the final output of the risk diagram gives the level of SIL of the system under scrutiny (1, 2, 3 or 4) and corresponds to a measure of risk reduction required for the system.

Due to the possibility of adapting the parameters to the various fields and professions of the company, the main challenge resides in the calibration of the graph. The calibration of the parameters must take into account all the possible situations.

One of the difficulties is to take into account the lack of precision and uncertainty on the part of the experts. In order to address this, approaches based on fuzzy logic have been proposed; see e.g. [SIM 06].

## 4.4.2. LOPA

As indicated in the IEC 61511 standard, the layer of protection analysis (LOPA) method (see [AIC 93]) starts with processed data from the HAZOP analysis and takes into account each of the identified hazards by documenting the initial cause and protective layers that help avoid or mitigate hazards. As opposed to purely qualitative techniques of risk assessment, the analysis of protective layers provides an estimation of the frequency for a given feared event.

The LOPA method evaluates how to reduce risk by analyzing the contribution of the different layers (intrinsic characteristics of the process up to the rescue methods) in the event of an accident. It is used to determine which SIL is assigned to each safety-instrumented function (SIF) and how many protective layers are required to bring risk to a tolerable level. The aim is to calculate residual risk, expressed in accident frequency per year, which necessitates quantification of the frequency of occurrence of initiating events, as well as the failure probabilities of each layer.

The analysis is divided up into the four following steps:

– defining the impact of the feared event (seriousness);

– determining and listing all the possible initiating events;

– determining and listing all the protective layers that prevent the initiating event leading to the feared event and from spreading;

– determining the frequency of initiating events, based on data or experts' judgments;

– determining the effectiveness of the protective layers in terms of the probability of on-demand failure;

– calculating the frequency of feared events.

The LOPA method only applies to on-demand functions (i.e. where the safe system is only brought into use in an event initiating a dangerous situation, which is independent of it) and it is not adapted to the continuous mode (a failure in the security system is initiating the dangerous situation).

As opposed to the risk diagram method, the LOPA method does not require calibration, as the inputs are finite quantities. The problem still arises as to the values, as they are not commonly accepted and differ according to sites, situations, feedback, etc.

### 4.4.3. *Summary*

One of the main challenges introduced by this type of method is the necessity to validate a number of parameters, as these parameters have a direct effect on the choice of SIL.

The use, in railway applications, of products whose safety levels are assigned to this type of approach will be faced by the difficulty of proving that the chosen hypotheses are verified by their context of use.

### 4.5. Conclusions

In this chapter, we have presented the basic methods for SIL allocation, such as it is applied in the context of railway signals.

As a general rule, the same approach is used for other rail subsystems, but standards such as IEC 61508 [IEC 08] are increasingly applied to appendices (ventilation management system in a tunnel, voiced announcing system, etc.).

We have aimed to define standards, for a rail system, as a set of products. A rail system could be viewed as a LEGO construction but with the need to verify the compatibility of its products.

Accepting previously certified products is thus a major issue that requires cross-acceptance analyses.

### 4.6. Bibliography

[AFN 01] AFNOR, EN 50155, French standard, Applications Ferroviaires. Equipements électroniques utilisés sur le matériel roulant, December 2001.

[AHM 09] AHMED A., LANTERNIER B., "Trois méthodes pour déterminer le niveau de sécurité", *Revue Mesure*, Numéro 813, pages 26-27, 2009.

[AIC 93] American Institute of Chemical Engineers, *Guidelines for Safe Automation of Chemical Processes*, CCPS, New York, 1993.

[BAR 90] BARANOWSKI F., Définition des objectifs de sécurité dans les transports terrestres, Technical Report 133, INRETS-ESTAS, 1990.

[BAU 10] BAUFRETON P., BLANQUART J.P., BOULANGER J.L., DELSENY H., DERRIEN J.C., GASSINO J., LADIER G., LEDINOT E., LEEMAN M., QUÉRÉ P., RICQUE B., "Multi-domain comparison of safety standards", *ERTS2*, 19-21 May 2010, Toulouse France.

[BAU 11] BAUFRETON P., BLANQUART J.-P., BOULANGER J.-L., DELSENY H., DERRIEN J.-C., GASSINO J., LADIER G., LEDINOT E., LEEMAN M., QUERE P., RICQUE B., "Comparaison de normes de sécurité-innocuité de plusieurs domaines industriels", *Revue REE*, No 2 – 2011.

[BIE 98] BIED-CHARRETON D., Sécurité intrinsèque et sécurité probabiliste dans les transports terrestres. Technical Report 31, INRETS - ESTAS, November 1998.

[BLA 08] BLAS A. and BOULANGER J.-P., "Comment Améliorer les Méthodes d'Analyse de Risques et l'Allocation des THR, SIL et Autres Objectifs de Sécurité", *LambdaMu 16, 16ème Congrès de Maîtrise des Risques et de Sûreté de Fonctionnement*, Avignon 6-10 October 2008.

[BOU 00] BOULANGER J.-L. and GALLARDO M., "Processus de validation basée sur la notion de propriété", *LamnbdaMu 12*, 28-30 March 2000.

[BOU 06] BOULANGER J.-L., Expression et validation des propriétés de sécurité logique et physique pour les systèmes informatiques critiques, Compiègne University of Technology, May 2006.

[BOU 07] BOULANGER J.-P. and SCHÖN W., "Assessment of Safety Railway Application", *ESREL 2007*.

[BOU 06b] BOULANGER J.-L. and SCHÖN W., "Logiciel sûr et fiable : retours d'expérience", *Revue Génie Logiciel*, December 2006, No 79, pp 37 - 40.

[BOU 08] BOULANGER J.-P. and GALLARDO M., "Poste de manœuvre à enclenchement informatique: démonstration de la sécurité", *CIFA, Conférence Internationale Francophone d'Automatique*, Bucharest, Romania, November 2008.

[BOU 09] BOULANGER J.-L., *Sécurisation des architectures informatiques – exemples concrets*, Hermès-Lavoisier, Paris, France, 2009.

[BOU 09b] BOULANGER J.-L., "Le domaine ferroviaire, les produits et la certification", *Journée « ligne produit" 15 October 2009*, Ecole des mines de Nantes.

[BOU 10] BOULANGER J.-L., "Sécurisation des systèmes mécatroniques. Partie 1", dossier BM 8070, *Revue technique de l'ingénieur*, November 2010.

[BOU 11] BOULANGER J.-L., *Sécurisation des architectures informatiques industrielles*, Hermès-Lavoisier, Paris, France, 2011.

[BOU 11a] BOULANGER J.-L., "Sécurisation des systèmes mécatroniques. Partie 2", dossier BM 8071, *Revue technique de l'ingénieur*, April 2011.

[BOU 99] BOULANGER J.-P., DELEBARRE V. and NATKIN S., "Meteor: Validation de Spécification par modèle formel", *Revue RTS*, no. 63, pp47-62, April - June 1999.

[CEN 99] CENELEC, NF EN 50126, Railway applications - The specification and demonstration of Reliability, Availability, Maintainability and Safety (RAMS), October 1999.

[CEN 01a] CENELEC, NF EN 50128, Applications Ferroviaires. Système de signalisation, de télécommunication et de traitement – Logiciel pour système de commande et de protection ferroviaire, July 2001.

[CEN 01b] CENELEC, EN 50159-1, European standard, Applications aux Chemins de fer : Systèmes de signalisation, de télécommunication et de traitement - Partie 1 : communication de sécurité sur des systèmes de transmission fermés, March 2001.

[CEN 01c] CENELEC, EN 50159-2, European standard, Applications aux Chemins de fer : Systèmes de signalisation, de télécommunication et de traitement - Partie 2: communication de sécurité sur des systèmes de transmission ouverts, March 2001.

[CEN 03] CENELEC, NF EN 50129, Railway applications – Communication, signalling and processing systems – Safety related electronic systems for signalling, February 2003.

[CEN 11] CENELEC, NF EN 50128, Applications Ferroviaires. Système de signalisation, de télécommunication et de traitement – Logiciel pour système de commande et de protection ferroviaire, July 2011.

[CEN 07] CENELEC, European standard, Railway Applications – Communication, Signalling and Processing systems – Application Guide for EN 50129 – Part 1: cross-Acceptance, May 2007.

[DG 06] DG ÉNERGIE ET TRANSPORT. Ertms - pour un trafic ferroviaire fluide et sûr. Technical report, European Commission, 2006.

[GUL 04] GULLAND W G, "Methods of Determining Safety Integrity Level (SIL) Requirements - Pros and Cons", *Proceedings of the Safety-Critical Systems Symposium*, Birmingham, UK, February 2004.

[HAD 95] HADJ-MABROUCK H., "La maitrise des risques dans le domaine des automatismes des systèmes de transport guidés: le problème de l'évaluation des analyses préliminaires de risqué", *Revue Recherche Transports Sécurité*, (49):101–pp. 112, December 1995.

[HAD 98] HADJ-MABROUCK H., STUPARU A., and BIED-CHARRETON D., "Exemple de typologie d'accidents dans le domaine des transports guides", *Revue générale des chemins de fers*, 1998.

[HAR 06] HARTWIG K. GRIMM M., MEYER ZU HÖRSTE M., *Tool for the Allocation of Safety Integrity Levels, Level Crossing*, Montreal, 2006

[IEC 91] IEC, IEC 1069: Mesure et commande dans les processus industriels - appréciation des propriétés d'un système en vue de son évolution, Technical report, 1991.

[IEC 08] IEC, IEC 61508: Sécurité fonctionnelle des systèmes électriques électroniques programmables relatifs à la sécurité, international standard, 2008.

[IEC 05] NF EN 61511, European standard, Sécurité fonctionnelle, Systèmes instrumentés de sécurité pour le secteur des industries de transformation, March 2005.

[ISA 05] ISA, Guide d'Interprétation et d'Application de la Norme IEC 61508 et des Normes Dérivees IEC 61511 (ISA S84.01) ET IEC 62061, April 2005.

[ISO 08] ISO, ISO 9001:2008, Systèmes de management de la qualité - Exigence, December 2000.

[ISO 09] ISO, ISO/DIS26262, Road vehicles – Functional safety – not published, 2009.

[LAP 92] LAPRIE J.C., AVIZIENIS A., KOPETZ H. (eds), *Dependability: Basic Concepts and Terminology, Dependable Computing and Fault-Tolerant System*, vol. 5, Springer, New York, 1992.

[LIS 95] LIS, Laboratoire d'Ingénierie de la Sûreté de Fonctionnement. *Guide de la sûreté de fonctionnement*, Cépaduès, 1995.

[LEV 95] LEVESON N.G., *Safeware: System Safety and Computers*, Addison Wesley, Nenlo Park, CA, USA, First ed. 1995.

[PAP 10] PAPADOPOULOS Y., WALKER M., REISER M.-O., WEBER M., CHEN D., TÖRNGREN M., SERVAT D., ABELE A., STAPPERT F., LONN H., BERNTSSON L., JOHANSSON R., TAGLIABO F., TORCHIARO S., SANDBERG A., "Automatic Allocation of Safety Integrity Levels", *Workshop CARS, Congrès EDCC 2010*.

[RSS 07] RAIL SAFETY AND STANDARDS BOARD, *Engineering Safety Management (The Yellow Book) – Fundamentals and Guidance*, Volumes 1 and 2, Issue 4, 2007.

[SIM 06] SIMON C., SALLAK M. et AUBRY J.F., "Allocation de SIL par Agregation d'Avis d'Experts/ SIL Allocation by Aggregation of Expert Opinions", *LambdaMu 15*, 2006.

[SMI 07], SMITH D.J. and SIMPSON K.G.L., *Functional Safety, a Straightforward Guide to Applying IEC 61508 and Related Standards*, second edition, Elsevier, 2007.

[STR 06] STRMTG, Mission de l'Expert ou Organisme Qualifié Agrée (EOQA) pour l'évaluation de la sécurité des projets, version 1 du 27/03/2006.

[VIL 88] VILLEMEUR A., *Sûreté de fonctionnement des systèmes industriels*, Eyrolles, Paris, 1988.

## 4.7. Acronyms

ASIL:                    Automotive SIL

CENELEC[11]:        *Comité Européen de Normalisation Électrotechnique* (European Committe for Electrotechnical Standardization)

DAL:                    Design Assurance Level

EMC:                    ElectroMagnetic Compatibility

EOQA:                  *Expert ou Organisme Qualifié Agrée* (Agreed Qualified Expert or Body)

---

11 See www.cenelec.eu.

| FR: | Failure Rate |
| GSM-R: | Global System for Mobile communications - Railways |
| HL: | HazardLog |
| IEC[12]: | International Electrotechnical Commission |
| LOPA: | Layer of Protection Analysis |
| OS: | Operating System |
| RSD: | Register of Dangerous Situation |
| SC: | Safety Case |
| SIF: | Safety-Instrumented Function |
| SIL: | Safety Integrity Level |
| SSIL: | Software SIL |
| THR: | Tolerable Hazard Rate |

---

12 To learn more, visit: http://www.iec.ch.

# 5

## Principles of Hardware Safety

The safety of an electronic computing unit is based upon architecture management. The hardware architecture can use several techniques such as partitioning, redundancy and/or diversity to achieve the safety objectives (tolerable hazard rate, design assurance level, etc.).

### 5.1. Introduction

This chapter[1] aims to present all the various techniques that allow a safe functioning of the hardware. We shall concentrate on the hardware as safety can rely on one or several electronic computing units (ECU). We shall leave "software" aspects out along with design.

In this chapter we just discuss the safety of hardware architecture, the objective (THR, SIL, etc), the process and the safety demonstration are discussed in other chapters (see Chapters 4 and 7).

### 5.2. Safe and/or available hardware

At the hardware architecture level, the main type of failure results from faulty output. There are two possibilities:

– an erroneously permissive output resulting in a safety problem (e.g. a traffic light turned to green mistakenly allows a vehicle to pass);

---

1 This chapter is based on educational material produced together with M. Walter SCHÖN, Professor at the University of Technology of Compiègne, whom I can never thank enough.

– an erroneously restrictive output resulting in availability problems (e.g. the train is stopping).

We can distinguish two categories of systems according to the impact of the absence of any output:

– integrity systems: should not have any erroneous outputs (incorrect data or correct data at an incorrect time, etc.). Integrity systems are those systems for which the process is irreversible (e.g. banking transactions). For this kind of system, it is preferable for all functions to stop rather than to malfunction. The system is called *fail-silent* (*fail-safe*, *fail-stop*);

– persistent systems: do not dispose of a fall-back state, which means that the absence of data may lead to a loss of control. For this type of system, it is more desirable to obtain some incorrect data rather than none at all. The system is said to be *fail-operate*.

An integrity system becomes safe if it can reach a fall-back state passively. For instance, in the railway sector, any failure results in cutting the power supply. And without power, braking occurs. The train reaches a safe state passively: "train stopped".

## 5.3. Reset of a processing unit

Section 5.2 allowed us to introduce the issue of the necessity, whether present or absent, to dispose of a fall-back position within the discussion on persistency and integrity.

In the case of integrity equipment, the transition to a fall-back state is irreversible. As for a transient fault, the resulting unavailability of the system may be unacceptable from the point of view of the client (e.g. the loss of the ABS function in a car). This is why it is tempting to go through an intermediary stage, which is the attempt to reset all or part of the piece of equipment (a treatment unit among *n* treatment units).

The use of the reset option must be controlled as several problems may occur:

– reset of a processing unit upon a failure, may lead to the reset of the requesting unit through a divergence of contexts, and the risk is to enter a reset loop that gets out of hand. The outputs then have to be made sure to be in a restrictive state during those intermediate states;

– the reset time can be lower than the detection time for the error and, in spite of reset requests, the system then produces outputs even though there is an error. A *reset loop* is detected through the use of a reset counter which has to be kept under control. Moreover, the resetting must be shown to have an effect on the particular failures that are being covered;

– etc.

When it comes to resetting the equipment, attention must be paid to demonstrating that the measure is effective and that there is no risk of masking a faulty situation.

## 5.4. Presentation of safety control techniques

The safety of the hardware part can be achieved through five main techniques:

– error detection (section 5.4.1);

– diversification (section 5.4.2);

– redundancy (section 5.4.3);

– retrieval (section 5.4.4);

– partitioning (section 5.4.5).

In this section, we shall go into more detail on these techniques and discuss their use.

### 5.4.1. *Error detection*

5.4.1.1. *Concepts*

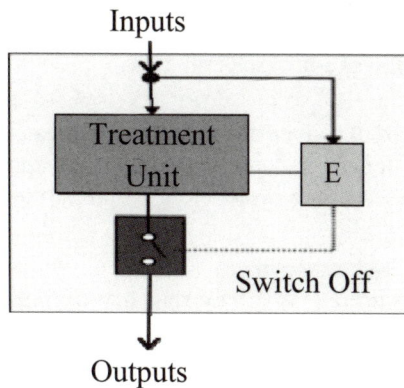

**Figure 5.1.** *Principles of error detection*

As shown in Figure 5.1, this technique aims to complement the hardware with an element that detects errors. So, in the event of an error, various solutions can be envisaged such as restarting, or switching off the outputs. The architecture in Figure 5.1 is integrated; in the event of error detection, the outputs are switched off.

The correct implementation of error detection depends on three techniques:

– detection of time consistency: this is achieved through the use of a "watchdog" which can identify the time interval error of the application, infinite loops or the failure to observe a deadline. The watchdog can be either a hardware or software element;

– detection of a hardware error: is achieved through self-tests. These self-tests help testing, more or less completely, hardware elements (ALU[2], memory unit processing, memory, voter, etc.). They can unfold entirely or partly during initialization, at the end of the mission, for some or every period. The main challenge with this technique resides in the relevance of the tests (coverage and completeness) according to the time of their execution;

– detection of an execution error: this is achieved through verification of the consistency in the application behavior. It is possible to obtain an analysis of the input consistency (input correlation, redundancy, etc.), or the output consistency (a given output cannot evolve randomly from one cycle to the next), behavior consistency (management of the "flag" within the code allows us to verify the execution path). It is possible to further correct the execution path (the "offline" calculation of all the execution paths as well as its "online" equivalent).

### 5.4.1.2. *Watchdog*

Simple and cheap, watchdogs are used to detect errors (that are frequent and not always acceptable) that lead to the inactivity ("crash") of a central unit. More broadly, it allows detection of a time interval error.

The time interval error of a given architecture can be caused by different types of failures: failure of the timers, failure of the software application (infinite loop, blocking, exceeding the treatment cycle, etc.), failure at the level of the resources, etc.

In Figure 5.2, the watchdog is a hardware device, regularly refreshed by the processor (e.g. at the beginning of each application lifecycle) and which can switch the outputs. It is to be noted that the watchdog can also be a software application.

---

2 ALU stands for Arithmetic and Logic Unit, a part of the processor.

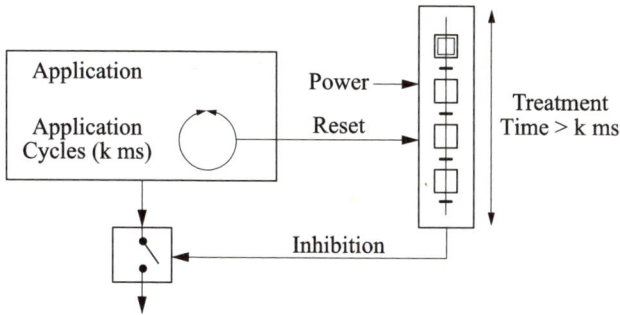

**Figure 5.2.** *Watchdog*

There is no general rule; however if the refresh does not occur after a set amount of time, the watchdog sends an error signal which can be used for example to:

– reset the processor;

– inhibit the outputs: *fail silent* system;

– change a unit: use of a different unit instead;

– etc.

### 5.4.1.3. *Self-tests*

5.4.1.3.1. Presentation

| Element | Test Type |
|---|---|
| ALU and/or instruction set | Specific tests allowing detection of errors in the instructions and their addressing modes |
| RAM[3] | Writing/reading test, e.g. the 55/AA type |
| ROM, FLASH | Checksum or 16 bit CRC[4] protection |
| EEPROM | Checksum or 16 bit CRC protection |
| I/O | Hardware provided to test inputs and outputs |
| Horloge/*Timer* | Exchange and comparison of signals between two microcontrollers each with its own clock |

**Table 5.1.** *Families of self-tests*

3 RAM, ROM, FLASH and xPROM are different kinds of memory used.
4 The notion of CRC (cyclic redundant code) will be explained later, see section 5.4.3.2.4.

After detecting a time interval error, it is possible to identify a second source of failure, namely that related to "hardware" which means: memory failure, failure of the treatment units, etc.

In order to detect hardware failures, it is possible to create specific tests. These tests aim to check that the hardware material is able to perform the expected service. Table 5.1 presents the type of tests that can be associated with the different elements of the computer architecture.

### 5.4.1.3.2. Example

Table 5.1 succinctly presents the types of tests which can be applied to detect the various types of hardware failures. We shall use as an example the detection of RAM failures.

The simplest technique and most demanding in terms of treatment time is the implementation of a write and read-back test. This test can be random (the tested memory cell and/or the written value are not predetermined) or predefined. The writing test with proofreading can be performed on the whole or part of the memory storage.

This type of verification consists in saving the content of the memory storage, testing and performing the writing test with proofreading. In general, the test is performed by writing a value (e.g. 55 and/or AA, as these values should be complementary) and by checking through proofreading that the writing is successfully accomplished. The main interest resides in the simplicity and effectiveness of the technique; however a link exists in the relation between the memory size and the operation time.

A variation of this test consists in choosing a figure (circle, square, cross, etc.), in storing it in the memory and calculating the checksum of the entire storage. There are considerable benefits in terms of treatment time and it adds to the failure detection of the ALU which is used to work out the checksum.

Finally, it may be necessary to focus checks on a set of data considered "critical" for the system in question. It is then possible to write each critical datum in two zones of the RAM (different zones with two different memory banks, etc.) and to check consistency between the two copies of the datum. The copies may be identical or not. For instance, if we choose to write the value and its complement, consistency can be verified by summing them up and comparing them with 0. This technique allows us to improve the execution time for the control and uses the ALU, but it has a major disadvantage in its high consumption of memory space.

5.4.1.3.3. Strategies of implementation

These tests (called self-tests) can be run at different stages:

– the initialization stage: the objective is to check that the system is able to perform its assigned mission;

– the end of the mission: the aim is to perform a check-up of the system;

– cyclically: the objective is to detect potential failures during the mission.

The cyclic performance of the tests requires suspending any other treatments, including the system's main task, or the saving of the contexts (all the memory used).

In any case, this mechanism can only detect temporary failures if they appear during the testing period.

The main difficulties in carrying out these tests lie in the selection of the packages so as to give a good coverage of the failures and to optimize the frequency of the tests.

5.4.1.4. *Consistency checks*

Consistency checks do not aim to detect all the existing failures but check that the general state of the system is consistent with a specific criterion. To conclude, they do not check the correction. It is possible to detect errors by checking:

– consistency among inputs;

– consistency between outputs and inputs;

– consistency in the execution by creating check points allowing us to verify traces;

– etc.

It would be impossible to present all the different types of checks; this is why we shall only focus on three types.

5.4.1.4.1. Consistency among inputs

Consistency among inputs relies on the fact that there is redundancy among the inputs overall:

– between two distinct acquisitions, e.g. within the same chain or two different chains;

– a given acquisition is unique but with different sources for the final information, e.g. speed is measured through two wheels with different gears;

– a piece of information may be coded (the notion of code introduces redundancy);

– inputs may be contradictory (two opposing values);

– etc.

5.4.1.4.2. Consistency between outputs and inputs

For some systems, post-conditions make it possible to check consistency in the treatment performed. These post-conditions link outputs to inputs.

Some of these post-conditions address a particular aspect (e.g. measuring an angle) and are related to issues of physics (maximum acceleration), to implementation choices (the last element of a list is always the null element), etc.

For example when measuring the angle of a steering wheel in a car, this angle cannot physically change by more than 90° over the cycle.

5.4.1.4.3. Consistency in the execution

Consistency checks in the execution help in assessing whether a given software application is following a previously validated path. In order to do so, the execution of the software application must be traceable. The execution is composed of a number of *traces,* and each trace is a series of checkpoints (execution path).

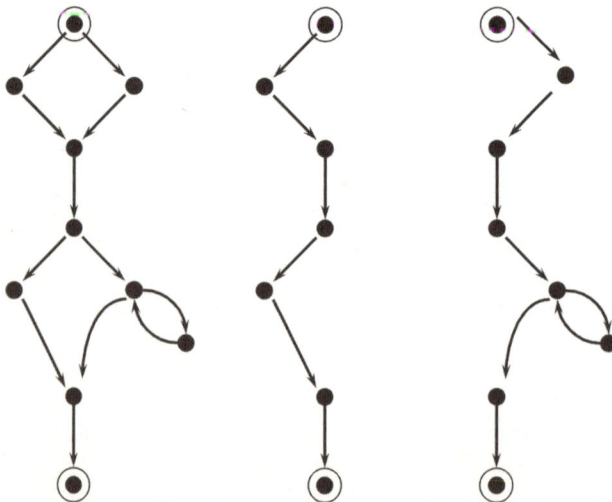

**Figure 5.3.** *Execution traces*

The first graph of Figure 5.3 represents a program with two consecutive IF instructions and a WHILE instruction. Analysis of the operating execution paths can lead to the selection of two execution traces. These traces are characterized by the creation of checkpoints. These indicators can be local (several pieces of information are stored within them) or general (only one variable is handled).

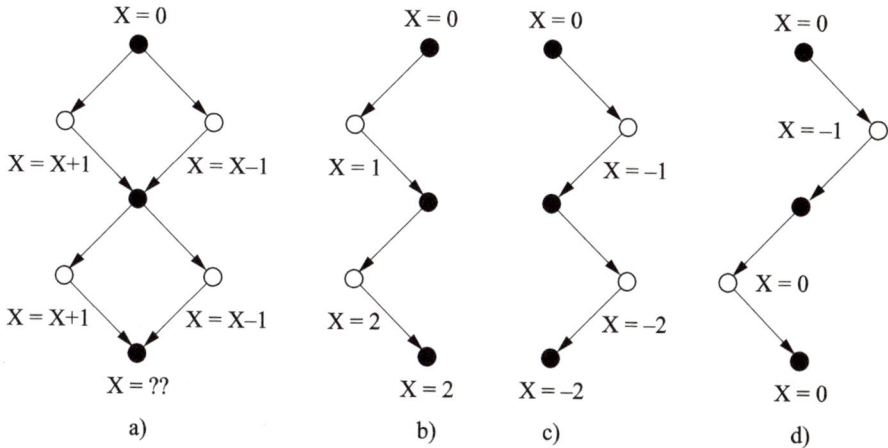

**Figure 5.4.** *Example of flag calculation*

As an example (Figure 5.4), it is possible to have a variable starting at 0, which, when it switches to a THEN branch, is increased by 1; when it switches to ELSE it is decreased by 1; and which, by the end of the execution must be equal to 2 or -2.

In some systems with high levels of criticality, the checkpoints are in high numbers and help in controlling the final stage of the execution, or even the entire execution.

The more numerous and/or complex these traces are, the higher the number of checkpoints. This has consequences on the memory space (storage of the traces and of the current trace), on the execution time (complementary treatments), as well as on the complexity of the code (the addition of treatment unrelated to functional aspects, makes the code and the analyses associated with it more complex).

### 5.4.1.5. *Summary*

The mechanisms for error detection are relatively easy to put into place and help us identify failures at the hardware level which can result from the behavior of the equipment (hardware + software).

Error detection helps make the execution of applications safe from certain types of failures (time interval errors, tool failures, treatment failures).

It can help in building a so-called *fail silent* system, i.e. with no outputs issued when errors are detected, but it also creates the basis for error tolerance, as, in the event of anomaly detection, it is possible to switch the calculator when there is redundancy.

### 5.4.2. *Diversification*

Diversification consists in creating a given component (i.e. a given function) in several distinct ways. The idea is that these components are under different environmental constraints, so that failures are different for each of them too.

Among the diversification techniques, we find:

– ecological/architectural diversification: the elements (the arithmetic unit, the software, the operating system, etc.) are different; the errors that affect each element will then also be different;

– geographical diversification: the elements are stationed in different places, so that the environmental constraints will be different;

– spatial diversification: different versions of the same software are used on different machines. The idea is that each version of the software will have its own flaws. This technique is also used to implement commercial off-the-shelf (COTS), and different versions of COTS are used;

– temporal diversification: creation of a regular development of the application to be executed (development of the deadlines, of the settings, and/or of the application);

– modal diversification: the system employs several methods to acquire the same data (network, modem, wired link, satellite access, etc.), for power supply, etc.

Ecological/architectural diversification is the most common form of diversification. When it is applied to the software, it is important to show that there actually is architecture diversification, thereby avoiding repetition of flaws within different software packages.

### 5.4.3. *Redundancy*

Redundancy aims to multiply a given resource more times than necessary when it functions correctly. Redundancy is generally divided into three parameters:

– time: more time than is necessary is employed to carry out the treatment. The application will be used at least twice on the same arithmetic unit. This simple technique requires a reference to compare (vote) the results (self-tests, etc.);

– information: there is more data than is necessary, so that the encoding of the information (parity digit, sum checks, cyclic redundancy control, Hamming code, etc.). This encoding can be used to detect errors but also to correct them;

– hardware: there is more equipment than necessary. Equipment redundancy is the basic technique which allows the creation of nOOm architecture (*m* bigger than *n*). Among these types of architecture are 2oo2 or 2oo3 (*x* treatment units do the calculation and a voter helps checking whether the results are correct). Architecture 2oo2 improves assurance, whereas architecture 2oo3 improves assurance and is available. Hardware redundancy can be passive (serves as backup equipment) or active. It would not be possible to list all the types of noom architecture in this chapter, so we shall only focus on a few representative examples.

The main interest in redundancy lies in that it helps detecting random failures that occur punctually. As with systematic failures, redundancy only helps detecting flaws in software applications (e.g. an arithmetic unit that cannot perform sums). This is why redundancy is generally backed up through the use of diversification.

### 5.4.3.1. *Time Redundancy*

#### 5.4.3.1.1. Presentation

Time redundancy uses twice the same application through the same treatment unit (processor, etc.). The results are usually compared through a device external to the processor and any inconsistency provokes a fall-back of the arithmetic unit (*fail stop* behavior). This technique is often used for programmable automatons.

Figure 5.5 shows a timeline for operational redundancy. The triplet "acquisition, execution, saving" appears twice and, in the end, a comparison of the saved results is made.

The first instance of operational redundancy can help detecting memory failures. A unique program is loaded in two different zones of the storage medium (two different zones in the addressing memory, two different memory supports, etc.). Memory failures (RAM, ROM, EPROM, etc.) can then be detected along with intermittent failures of treatment units.

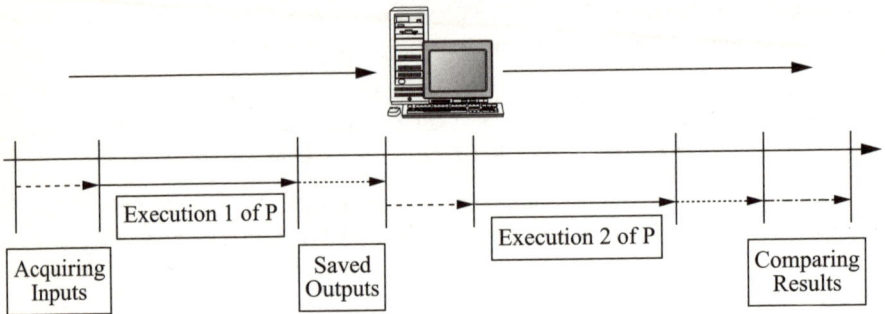

**Figure 5.5.** *Principles of operating redundancy*

It is to be noted that some failures of the shared hardware devices (comparison unit, treatment unit) are not detected and therefore stay latent. There are two ways in which errors can be masked:

– the outcome of the comparison of results can always be positive, no matter what the inputs are (failure of the comparison means is a common mode of failure);

– subtle failures (e.g. A-B is systematically performed instead of A+B) in the ALU part in the treatment unit can give the same erroneous result for each execution (the treatment unit is a common mode of failure).

One solution is to introduce self-tests along the execution; e.g. a comparison of data inconsistency (usually without the need to enter a fall-back state), comprehensive functional tests for the treatment unit. If we want effective error detection, the coverage of the tests must be wide enough (cover the usage instructions, etc.) and the tests must be performed at the appropriate moment (initialization, with each cycle, regularly, at the end of a mission, etc.). The major drawback with this solution, however, is its cost in terms of performance (related to the space taken by the self-tests as well as their frequency).

Another solution consists in introducing diversification of the code. This diversification can be "light" and we refer to it as voluntary dissymmetry of the encoding application. It is then possible to enforce the use of two different sets of instruction to program the application. For instance, one of the programs would use A+B, whereas the other would use -(-A-B).

A dissymmetry in the data can be introduced through addressing the stored object differently for each of the two programs (variables, constants, parameters, functions and procedures). It is to be noted that this voluntary dissymmetry can be introduced automatically within a single program. For these two types of dissymmetry the compilation phase must be controlled as well as the final executable to make sure that the dissymmetry is still present.

In general, redundancy should be complete (the entire application is executed twice), but a partial redundancy may be sufficient (Figure 5.6).

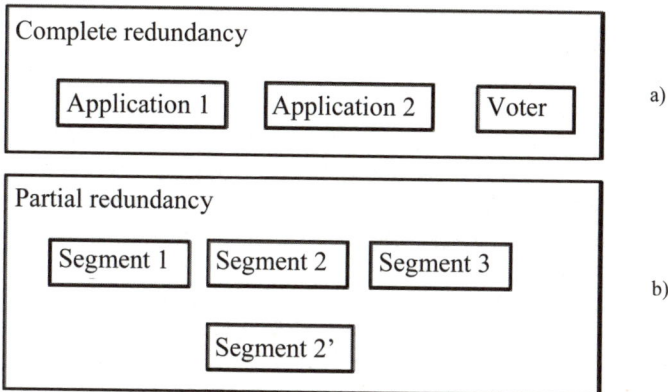

**Figure 5.6.** *a) Complete redundancy and b) partial redundancy*

The establishment of a floating-point computation in a safety function requires the implementation of a safety technique. The introduction of partial redundancy of the floating-point computation requires diversity, which might, for example, use different libraries. Therefore, it is necessary to accept certain errors during comparison, see for example Figure 5.7.

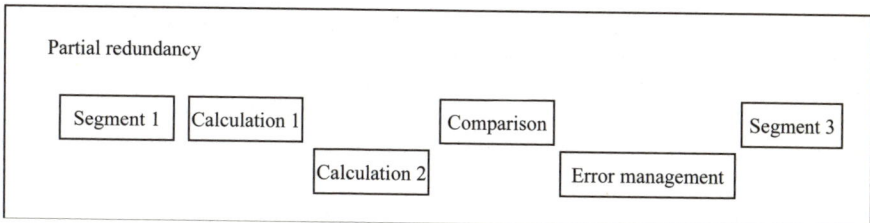

**Figure 5.7.** *Partial redundancy*

Software diversity can be accompanied by hardware diversity. As in [LEA 05], it is possible to have a hardware architecture offering the FPU (floating point unit) of the main processor and an annex unit to perform the computations. The diversification of the code is a little higher because we will use two sets of instructions. The concept of error acceptance here will be essential.

### 5.4.3.1.2. Example 1

Introducing partial dissymmetry is relatively simple but has only a weak error detection power. It is possible to generalize this solution through diversification of the application in question. There are various possibilities, e.g. using two teams, or two different code generators, etc.

As an example, Figure 5.8 shows how the EBICAB 900 works. The application being used is diversified into applications A and B. Application A is divided into F1, F2, F3; whilst application B is divided into F1', F2', F3'. Three development teams are then necessary; two independent teams are in charge of creating the two applications, whereas a third team is assigned the responsibility of specification (which is common to the two) and of synchronization. As there is only one acquisition phase, the data are protected (CRC, etc.). The data manipulated by application A are diversified (different memory, mirror bit by bit, etc.) compared to application B.

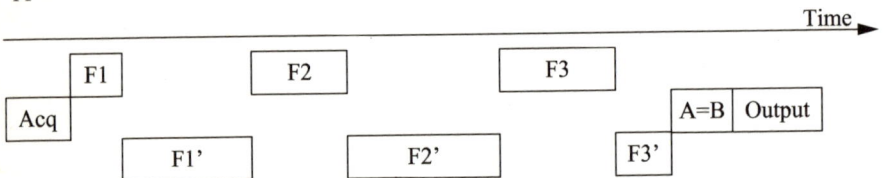

**Figure 5.8.** *Principles of Execution in Time EBICAB*

### 5.4.3.1.3. Example 2

As our second example, we will discuss a piece of field equipment that relates the "train target" (train targets are elements that give commands to trains). In this application, only one software application is formally developed with method B [ABR 96] and a treatment unit.

The B method is a formal method which ensures (through mathematical evidence) that the software is correct in terms of property. This warranty is interesting but does not cover the code generator, the chain generating the executable (compiler, linker, etc.) or the loader.

For this application there are two code generators and two interpreters executing the code (two compilers). This gives two different versions of the code and it has been shown that address tables (variables, constants, functions, parameters, etc.) for both executables are actually distinct. Each version of the applications is loaded in different memory spaces.

### 5.4.3.1.4. Summary

Time redundancy is a fairly simple technique which has the advantage of using a single treatment unit. Its main disadvantage on the other hand remains the long execution time it requires, as launching the double treatment with vote and self-test requires at least 2.5 to 3.5 times the time of a single treatment. This is why this type of solution is primarily used for systems whose treatment time is not an issue.

Partial or total diversification of the code allows a good error detection rate for both random and systematic errors (depending on the degree of diversification), although the cost will increase as there will be two software packages to maintain, etc.

### 5.4.3.2. *Information redundancy*

### 5.4.3.2.1. Presentation

Information redundancy is a widespread technique for error tolerance. It consists in manipulating an additional piece of information (so-called control or redundant) which helps detect, and even correcting the errors.

In this section, we shall consider:

– the parity check;

– the cyclic redundancy check;

– the Hamming code;

– the arithmetic check.

Generally information redundancy is used to detect failures related to stored data (memory, etc.) and/or transmitted data (data bus, network, etc.). However it can also be used, with more difficulty, to detect failures of the arithmetic units and or control units (self-checking).

As the communication network(s) of critical systems was (were) considered closed, most of the codes used are separable (control bits different from the bits used for the main information).

It is to be noted, however, that this point is changing with the implementation of the so-called "open" network (wireless, traveler network connected to the command network, etc.). In the future it will be necessary to use codes which are not separated (i.e. for which the extraction of the relevant part is more complex). As this section is an outline and not cryptography, we shall not go into further detail on this subject [LIN 99]).

### 5.4.3.2.2. Parity checks

A parity check is a very old technique which has long been used for modem transmission. It was the simplest error detection technique (called parity bit) and is added for each word. An additional bit is calculated so that the sum of the bits in the word (*modulo* 2) is equal to zero (even parity convention) or to one (odd parity convention).

Parity checks help in detecting an error (or several errors in even numbers). It does not detect double errors (or errors in odd numbers). In the event of error detection, there can be no correction.

Parity checks can be extended to the treatment of a whole group of words; in this case, it is referred to as a cross-parity check.

### 5.4.3.2.3. Checksums and process checks

With the executable, instead of adding a bit of information related to one or several words, it is common to have a piece of information called the checksum, which is the sum of the different words that form the application. The checksum can also be applied to a series of words, such as a column of the memory storage.

The checksum is easily calculated and it is possible to quickly check that the program loaded in the memory corresponds to what is expected. In the railway sector, the checksum is memorized in a plug that is read at the initialization phase. Any development of the application entails a change in the plug.

When it comes to networks, the information frames should be distinguished from the "lifeindicator" frames. Indeed, some systems may require the users to report for themselves through frames that do not have any functional needs. This is the case in the automotive sector, as some calculators need to emit frames that do not contain any information; but as a principle of information remanence applies, these frames have a relatively long lifetime on the network. A piece of information allows dating and hence excluding the use of these frames.

The use of these techniques does not pose any problem, but only covers a certain type of specific error.

5.4.3.2.4. Cyclic redundancy check

A cyclic redundancy check (CRC) is a powerful and easy to use means to control data integrity. It is the main method for error detection in the telecommunications field.

The principle behind cyclic redundancy checks consists in treating the binary as binary polynomials, i.e. the polynomials whose coefficients correspond to the binary sequence. So the binary sequence $M = u_{n-1}u_{n-2}\ldots\ u_1u_0$ can be considered as a polynomial $M(x) = u_{n-1}x^{n-1} + u_{n-2}x^{n-2} + \ldots u_1x + u_0$.

With this mechanism of error detection, a predefined polynomial (called generator polynomial), noted $G(x)$, is recognized by both emitter and receptor. The degree of $G(x)$ is k. Error detection implies for the emitter to run an algorithm on the bits of the frame so as to generate a cyclic redundancy check, and to transmit these two elements to the receptor. The receptor then only needs to do the same calculation in order to verify that the cyclic redundancy check is valid.

In practice, the message M corresponds to the bits in the frame to be sent and $M(x)$ the associated polynomial. We call M' (Figure 5.9) the transmitted message, i.e. the initial message to which the cyclic redundancy check of *k* bits was concatenated.

Cyclic redundancy checks such as $M'(x)/G(x) = 0$. The coded part of the check ($R(x)$) is equal to the rest of the polynomial division $x^k.M(x)$ ($M(x)$, to which are concatenated *k* null bits corresponding to the length of the CRC) times $G(x)$.

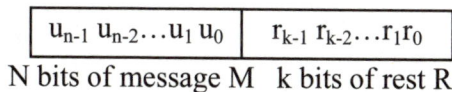

| $u_{n-1}\ u_{n-2}\ldots u_1\ u_0$ | $r_{k-1}\ r_{k-2}\ldots r_1 r_0$ |
|---|---|

N bits of message M   k bits of rest R

**Figure 5.9.** *Format of message M' with CRC*

The message $x^k.M(x) + R(x)$ is then emitted and the code becomes separable. When the message is received, if $M'(x)$ is not divisible by $G(x)$, it means that there is definitely an error. If $M'(x)$ is divisible by $G(x)$, it is highly probable (function of the choice of polynomial $G(x)$) that there is no error.

5.4.3.2.5. Hamming code

With the Hamming code, all the messages (vocabulary source over $k$ bits: usually made up of the $2^k$ possible words) are injected into a larger space ($n$ bits n > k), so that the coded words that correspond to messages are different enough.

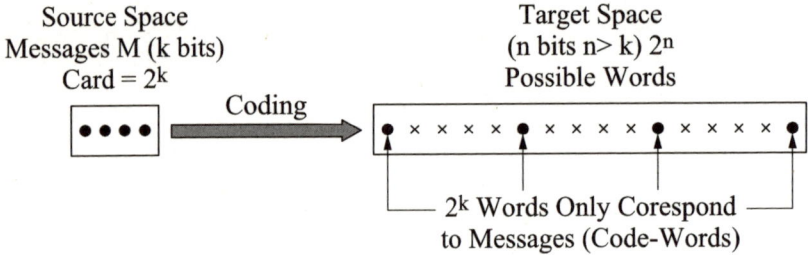

**Figure 5.10.** *Principles of the Hamming code*

The difference between the two coded messages (words corresponding to messages from the source vocabulary are said to be code-words) is measured as the Hamming distance, which is the number of bits that differ between them. The Hamming distance (noted *d*) of a code for a given source vocabulary (see Figure 5.11), is the value below the distance between two of the words in the code. On the other hand, words that do not refer to coded messages are said to be outside the code. A Hamming code is characterized by the triplet [k, n, d].

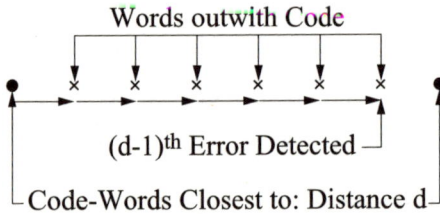

**Figure 5.11.** *Hamming distance*

A code with a Hamming distance d is able to detect a maximum of d-1 errors. Whereas a code with a Hamming distance of d = 2D + 1 is able to correct a maximum of D errors, by bringing the wrong word (outside the code) to the closest word within the code (see Figure 5.12).

In order to make the Hamming code more explicit, we will now describe a Hamming code characterized by the triplet [4, 7, 3]. The source messages take the form of $M_4M_3M_2M_1$, and the aim is to produce two check bits $C_1$, $C_2$, $C_3$ (which brings the targeted space up to seven bits) with a Hamming distance of three, which allows the detection of two errors and the correction of one.

The basic principle behind our code is as follows: $M_4$ intervenes in the calculation of 3 $C_i$ and each $M_i$ intervenes in the calculation of 2 $C_i$ different for the three.

The resulting message takes the form of $E_7\ E_6\ E_5\ E_4\ E_3\ E_2\ E_1 = M_4\ M_3\ M_2\ C_3\ M_1\ C_2\ C_1$. On the basis of the above choices and on the fact that number 7 is a 3-bit code, there is only one possible way of decomposing it with the three one-bit ones.

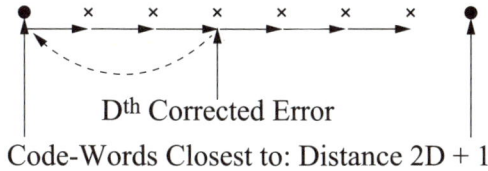

D^th Corrected Error

Code-Words Closest to: Distance 2D + 1

**Figure 5.12.** *Error correction*

Expressing the link between the elements of the message and the decomposition of position j of the associated $E_i$ element helps constructing the following equations:

– $M_4$ corresponds to $E_7$ where k = 111, so $M_4$ impacts on $C_1$, $C_2$, $C_3$;

– $M_3$ corresponds to $E_6$ where k = 110, so $M_3$ impacts on $C_2$, $C_3$;

– $M_2$ corresponds to $E_5$ where k = 101, so $M_2$ impacts on $C_1$, $C_3$;

– $C_3$ corresponds to $E_4$ where k = 100, so $C_3$ impacts on $C_3$;

– $M_1$ corresponds to $E_3$ where k = 011, so $M_1$ impacts on $C_1$, $C_2$;

– $C_2$ corresponds to $E_2$ where k = 010, so $C_2$ impacts on $C_2$;

– $C_1$ corresponds to $E_1$ where k = 001, so $C_1$ impacts on $C_1$.

We can then build $C_j$ through the following equations (+ for XOR):

– $C_1 = M_1 + M_2 + M_4$;

– $C_2 = M_1 + M_3 + M_4$;

– $C_3 = M_2 + M_3 + M_4$.

The Hamming distance is indeed three as two source messages, which differ by one $M_i$, result in two coded messages whose control part differs by at least two $C_j$.

Constructing the message is henceforth not a problem. The established code is separable (the position of the control elements is known). Once the message is received, calculation of $C_j$ with the $M_i$ helps revealing and correcting:

– if an equation is wrong, the erroneous bit is the $C_j$ bit in the equation in question;

– if two equations are wrong, the erroneous bit is the $M_i$ common to both equations;

– if all three equations are wrong, the erroneous bit is $M_4$.

On the basis of this principle, it is then possible to receive the message $m_7 m_6 m_5 m_4 m_3 m_2 m_1$, to calculate 3 $n_j$ such that $n_3 n_2 n_1$ reveals the erroneous bit in numbers. If it is 0, the transmission occurred correctly, or a simple error is detected and corrected and a simple error is detected but corrected inappropriately. The triple error may not be detected.

The $n_j$ are calculated through the following equations (+ for XOR):

– $n_j = m_1 + m_3 + m_5 + m_7$;

– $n_j = m_2 + m_3 + m_6 + m_7$;

– $n_j = m_4 + m_5 + m_6 + m_7$.

To conclude this section, once the Hamming code is constructed, it detects errors quickly and easily. The main challenge consists of working the code out (for a larger space) and in demonstrating that the Hamming distance is in fact that required for the entire development of the encoded application (adding variables, development of the functional fields, etc.).

## 5.4.3.2.6. Arithmetic code

The idea behind the arithmetic code ([FOR 89]) is to replace the data to coded data comprising of two fields: the value of the data (a part called "functional") and a coded part. The coded part contains information redundancy towards the functional part; and the latter's coherence is preserved through arithmetic operations.

The coded part requires the arithmetic operations to be replaced by specific operations that manipulate both parts.

| Value | Code |
|---|---|
| 16 x 5 = 80 | 2 x 5 [9] = 1 |

| 0 | 1 | 0 | 1 | 0 | 0 | 0 | 1 |
|---|---|---|---|---|---|---|---|

Total = 81

X = 81

| Value | Code |
|---|---|
| 16 x 7 = 112 | 2 x 7 [9] = 51 |

| 0 | 1 | 1 | 1 | 0 | 1 | 0 | 1 |
|---|---|---|---|---|---|---|---|

Total = 117

Y = 117

OPEL

Z = 81 + 117 = 198

| Total = 198 |
| z = 12 |

| 1 | 1 | 0 | 0 | 0 | 1 | 1 | 0 |
|---|---|---|---|---|---|---|---|

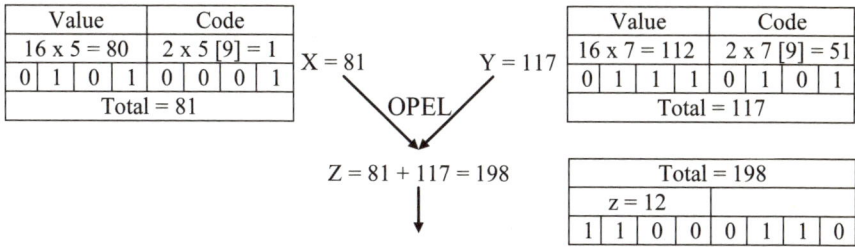

**Figure 5.13.** *Example of arithmetic code*

The Figure 5.13 present a simple example of an arithmetic code. We choose $A = 9$ and $k = 4$, we have $2^k = 16$ and $-2^k[A] = 2$. The data $X = 5$ is coded by 81, the data $Y = 7$ is coded by 117. The addition of X and Y gave 198. The code is separable and the decoding of Z (extraction of bit of high weight) gave $Z = 12$. The correctness is verified by the fact that $198[A] = 0$.

The coded part is constructed in such a way that it is possible to identify three types of errors:

– errors in the operations (and/or corruption) related to the data. In order to do so, a redundant piece of information is memorized as $-2^k.x$. This piece of information relies on the principle of the proof by nine, but nine is replaced here by A, which is a big number;

– treatment errors (executing an addition rather than a subtraction) through the implementation of a $B_x$. The $B_x$ is the signature associated to each variable X. This signature is predetermined (OFFLINE process) in a pseudo-random mode. For each variable resulting from an operation, it is possible to determine OFFLINE the expected $B_x$. Working out the $B_x$ must include the treatment of alternatives and of loops. The $B_x$ calculated OFFLINE are memorized in a PROM and the execution of the program with different input values always leads to the same signature development;

– errors resulting from the manipulation of obsolete data in the form of a date D.

The coding used has the next form $-2^k.x + B_x + D$ and is still separable.

The implementation of this particular coding and the operations associated with it allow one single processor to detect several types of errors. This is done at the price of a high computational cost induced by the checks performed (see [GUI 90]).

This technique [FOR 89, BIE 99, MAR 90] was developed to implement the SACEM[5] [GEO 90, MAR 90, HEN 94] in the 1980s, and its modes of use are detailed in the second chapter of [BOU 10]. It is used on many other applications such the SAET-METEOR ([MAT 98]).

5.4.3.2.7. Summary

The use of information redundancy is an interesting technique which helps detecting several families of errors, and even to make corrections. The examples used highlight the fact that the implementation of the code is more or less easy and can have an influence on the global execution time.

5.4.3.3. *Hardware redundancy*

5.4.3.3.1. Presentation

As shown in Figure 5.14, a redundant system is a set of treatment units which must be perceived as a single entity. In this chapter, it will not be possible to present all the possible combinations; however we will present the most significant ones.

**Figure 5.14.** *Hardware redundancy*

When it comes to the treatment units used, two alternative cases should be pointed out:

– either they are identical, and we refer to homogenous architecture;

– or they are different and we are in the presence of heterogeneous architecture; they are differentiated through diversification:

- the treatment units may be different,

---

5 The SACEM/HSDEM (in French "*système d'aide à la conduite, à l'exploitation et à la maintenance*" for help system to driving, exploitation and maintenance) is onboard the trains of line A of the Parisian RER.

- the treatment units may be identical with different operation systems and an identical application,

- the treatment units and the operation system are identical but the application differs,

- all the elements differ,

- etc.

As to redundancy, there are two possible forms:

– redundancy without a voter: one of the units is in charge of working out the outputs;

– redundancy with a voter: several units carry out the treatment and one additional element selects the outputs.

There is already a wide range of possibilities. In the following sections, we shall detail some of them; however this can only be a short overview.

5.4.3.3.2. Master/slave architecture

As shown in Figure 5.15, the master/slave architecture is made up of two identical treatment units. The "master" unit is responsible for all the treatments. It also triggers the sequences of activities of the so-called "slave" unit. It performs its treatments as well as the selection of outputs on the basis of its results and those of the "slave" unit.

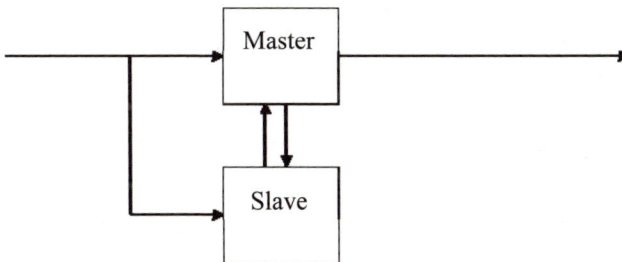

**Figure 5.15.** *Master/slave architecture*

One of the issues with this type of architecture consists in synchronizing the treatments. As the treatment units and the applications (*modulo* control of the master) are identical, a number of common failures (failing processor, flaw in the software, etc.) may go undetected.

As treatment units and applications are identical, the maintenance costs must be managed. The presence of a master unit ensures control over the outputs. The main disadvantage in this type of architecture resides in the absolute control of the master unit, as it is alone responsible for choosing the output values, without consulting with the slave unit.

This first type of architecture (Figure 5.15) can be complemented so as to improve the management of failures. Introducing a bilateral exchange pattern between the two treatment units; and assigning decision-making power (the triggering of reset to switch off the outputs, correction indicator, etc.) to the slave unit in case of divergence, will help in correcting the main flaw of the first type of architecture, as shown in Figure 5.16.

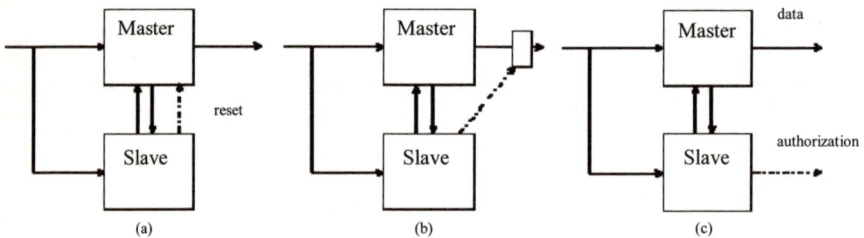

**Figure 5.16.** *Three alternatives*

The upgraded architecture remains sensitive to the common mode (hardware flaw, software flaw, etc.), but is better able to manage random failures.

In Figure 5.16 it can be said that case (b) is an integrity type of architecture and that case (c) is a persistent type. As for case (a), the slave unit can reinitialize the master unit. This introduces latency and its impact must be analyzed on the output.

The implementation of a double reset (each unit can trigger the reset of another one when detecting divergence) may introduce endless reset loops; see the discussion in section 5.3 on the reset of a treatment unit.

This type of architecture will, of course, be complemented by self-tests that ensure hardware flaw detection, protection of the data and of the program. The creation of a double acquisition chain helps detecting flaws at the acquisition stage. This type of architecture may be further complemented by dissymmetry of the application or diversification (in the hardware or software).

Figure 5.17 (see [BAL 03]) shows a type of architecture proposed by the automotive sector, where two identical treatment units perform the same treatment and share hardware elements (memory storage, bus, interrupt handler, etc.).

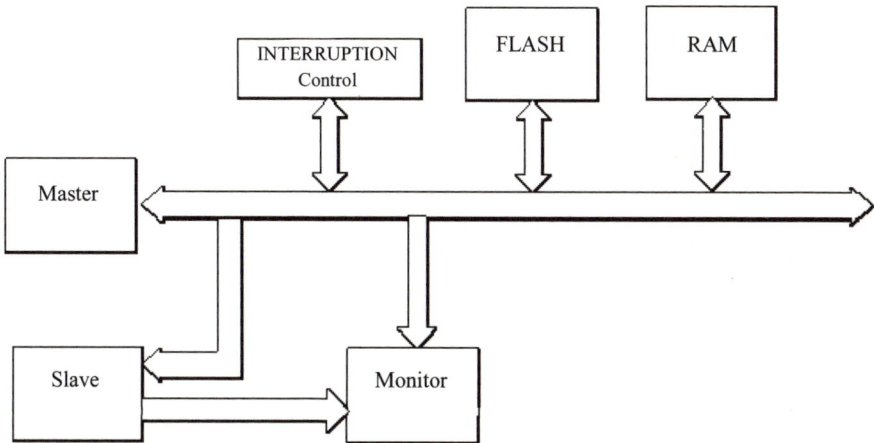

**Figure 5.17.** *Architecture lockstep dual processor*

Each instruction is sent to both treatment units and they each perform the treatment and make requests for the reading, writing and/or access to the next instruction. The "master" treatment unit manages the bus, whereas the slave unit is connected to a monitor. This monitor responds to the bus (thus to the master unit). In the event of a divergence when access is requested, or writing of a variable, or of access to the next instruction, the monitor triggers the safety mode.

The monitor has the responsibility of checking that the treatments performed by a given treatment unit concord. However it is not responsible for failures of the bus, memory or interrupt handler. The bus and memory may be protected through integrity checks. The interrupt handler, as it has a low level of complexity, will have to tolerate errors.

A variation of this type of architecture is the "orthogonal redundancy" [LEA 05]. It uses two different treatment units to perform the operations on the variables for example.

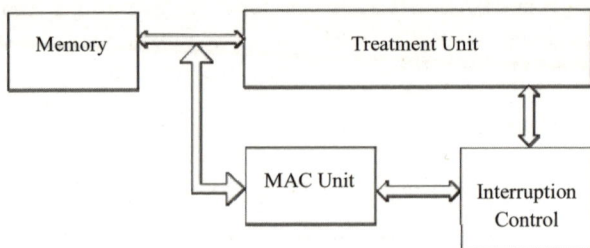

**Figure 5.18.** *"Orthogonal redundancy" architecture*

This type of architecture is based on a dissymmetry in the hardware resources (treatment units, memory storage, etc.). The main interest of this architecture is to have two different processor extension units (e.g. ALU and DSP[6]).

As an example, Figure 5.18 shows how the processor ST10 from STMicroelectronics is put into motion. This processor contains an ALU as an extension and can be interfaced to a MAC. The MAC (*Multiply and ACcumulate*) is a DSP. The ALU and the MAC each have a specific set of instructions for mathematical operations. The MAC optimizes operations such as multiplications, accumulation and digital filtering.

As shown in Figure 5.19, with the braking function, calculations on floating point variables are performed by the two processors extension units. The results are compared and should be equivalent, and this equivalence should be defined (precision issue and acceptability of the results).

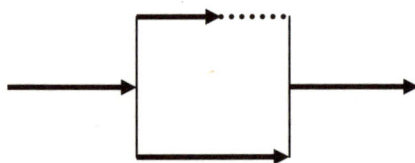

**Figure 5.19.** *Time aspect*

Orthogonal redundancy enables detection of failures at the ALU level for example when the floating point variables are manipulated.

---

6 DSP stands for Digital Signal Processor

5.4.3.3.3. Worker/checker architecture

A variation derived from the master/slave architecture consists in differentiating the role of each unit. One of the treatment units is dedicated to the treatment, whereas the second unit checks the running of the treatment.

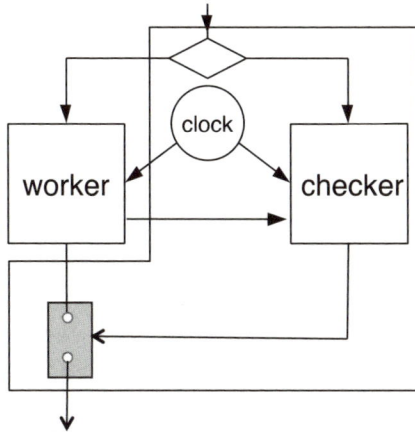

**Figure 5.20.** *Worker/checker*

As shown in Figure 5.20, a treatment unit (worker) works out the output; whereas a different treatment unit (checker) authorizes or not the output. The controller can perform the same treatment and compare through an approximate calculation or a verification of all the outputs produced by the first treatment unit. Neither the approximate calculation, nor output verification checks the correction, but only the plausibility of the results.

The treatment units need to be synchronized so as to facilitate a decision. This type of architecture enables diversification in terms of software applications. The higher the diversification, the weaker the impact of common failure modes entailed by the presence of identical hardware.

The maintenance costs for this type of architecture are slightly higher as there are two software applications to be managed. The main challenge then is to ensure that it will always be possible to decide on the plausibility of the calculated outputs during the lifetime of the system.

### 5.4.3.3.4. Architecture *NooK*

Architecture 2oo2 uses two identical treatment units and a voter (see Figure 5.21). Both units perform the same treatment and the results are then submitted to vote. If the results are identical, the output will be elaborated. If they diverge, on the other hand, the system may (a) stop (integrity system) or (b) point out the anomaly (persistent system).

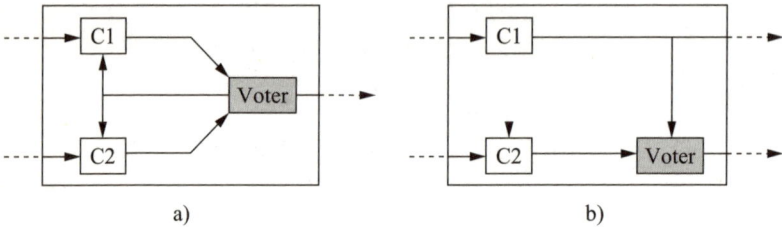

a)                                                              b)

**Figure 5.21.** *Architecture 2oo2*

This type of architecture helps detecting errors but is not tolerant of them. It is also to be noted that there are common modes of failure related to treatment units and to the software. As far as the software is concerned, it must be developed with an adequate level of safety. As to hardware flaws, self-tests will have to be put in place at least for the memory and processor.

The voter is a unique element; however the argument is that the hardware vote is a small-sized piece of equipment which can be analyzed thoroughly.

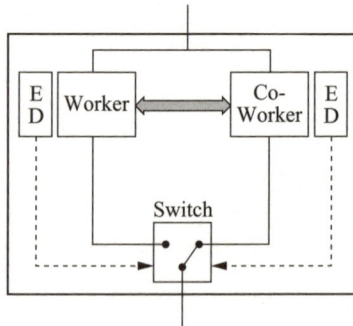

**Figure 5.22.** *1oo2D Architecture*

A variation of this 2oo2 architecture (see Figure 5.22) is the 1oo2D. Both treatment units are active but each is controlled by an error detection mechanism. These activate a switch which can then choose which outputs are authorized further.

Architecture 1oo2D is a persistent type of architecture. In the event of a failure of both treatment units (common mode or otherwise), the switch oscillates without being able to stabilize.

Architecture 2oo2 may be extended to three units, as shown in Figure 5.23, we are then referred to 2oo3.

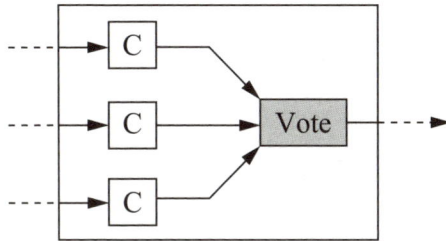

**Figure 5.23.** *Architecture 2oo3*

The treatment units simultaneously execute the same applications with either identical or diversified inputs. Their results are then compared by a voter (hardware and/or software) or by a pilot process. The results should be identical.

When a treatment cycle is performed, if the voter confirms that the results for all three units are similar, they are to be considered as functional. On the other hand, if the vote converges but one of the units is out of phase, it is considered subject to failure and removed from the cycle. In that case the architecture goes from 2oo3 to 2oo2. So the system tolerates one failure. But as the system now functions on 2oo2 architecture, the next failure will lead to a fall-back state.

In practice, because of synchronization problems, the results may differ; this is why the result, if it is a consensus, is considered correct.

A common physical clock may be used for synchronization, which then becomes a critical point in the architecture. The most common alternative solution consists in performing treatments asynchronously. However, this requires the software application to be resistant to delays, and for the voter to accept treating temporal windows.

In order to avoid problems of failure on the acquisition of the inputs and on the voter, triple modular redundancy (TMR) architecture triples the whole chain, as shown in Figure 5.24.

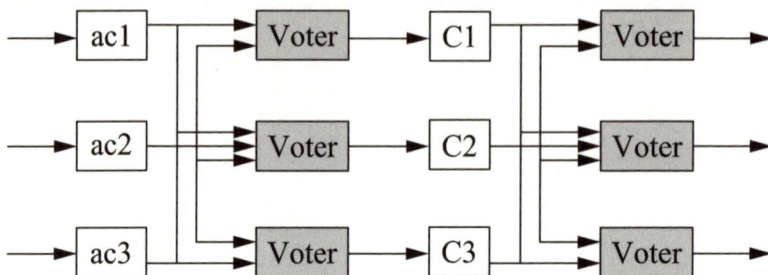

**Figure 5.24.** *Triple modular redundancy (TMR)*

With architecture 2oo3, having treatment units and the same software, leads to the presence of common mode failures.

This problem may be treated through controlling the quality of the software creation process, and selecting treatment units with a high feedback quality (the railway sector has used the processing units 68 020, 68 030 and 68 040 since the 1980s).

The behavior of new processors is still to be defined (anticipating operations, pseudo-parallelization, guessing the execution path, using an internal cache, using a multicore processor, etc.) and an efficient feedback mechanism still to be found. This explains the diversification of treatment units and/or of applications (diversification of sources, compilers, equipment, behavior in time, etc.).

As for the voter, it is to be noted that it is a delicate issue and that the current tendency goes towards using software voters. As much as managing a hardware voter is realistically feasible, software voters are a source of risk.

The main interest in having a software vote is its flexibility and the absence of a hardware complement. On the other hand, the execution time is increased.

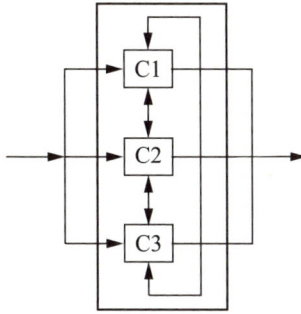

**Figure 5.25.** *2oo3 architecture with software vote*

As shown in Figure 5.25, when a software voter is used, it must come with a voting strategy. There are several possibilities of strategies:

– one of the treatment units is chosen as being definitely the master, which is not realistic in the event of failure in that particular unit;

– one of the units is elected with each cycle, according to a process to be defined;

– the voting cycle is predetermined;

– each unit performs a vote and produces a semi-output; so two semi-outputs are needed for an effective output;

– etc.

Architecture 2oo3 can be generalized to *NooK* architecture. Usually, K odd is used to perform the vote, even if, in some sectors such as the nuclear industry, we find K = 4 ([ESS 00], see section 3.2) or more.

The use of *NooK* enables the implementation of more complex software voting strategies such as:

– majority voting: with the *NooK* architecture, there needs to be $n$ identical values to obtain one vote;

– consensus vote: all $m$ results are compared and a score is associated to each of the values. The value with the highest score is chosen. If two values have the same highest score, it is not possible to make a choice;

– vote by taking confidence into account: the reliability of the software in voting is taken into account;

– etc.

The *NooK* architecture helps in isolating a failing unit and switching it off. This is generally a final action (principle of the fuse), but some types of architectures propose a learning mechanism which allows resetting the failing unit and teaching it the same context used by the other units:

– to give an example of *NooK of NooK* architecture, we shall briefly address its implementation for the Boeing 777. As shown in Figure 5.26 (for further information, see [YEH 96]), the overall primary flight computer (PFC) can be decomposed into three PFCs. The PFCs communicate through a network made up of three identical channels. A "Primary Flight Computer" is broken down into three lines, each self-powered and equipped with a calculator and communication system ARINC 629.

– each line in a PFC is provided with a different processor. These are: Intel® 80486, Motorola 68040 and AMD 29050. Each PFC uses all three processors differently.

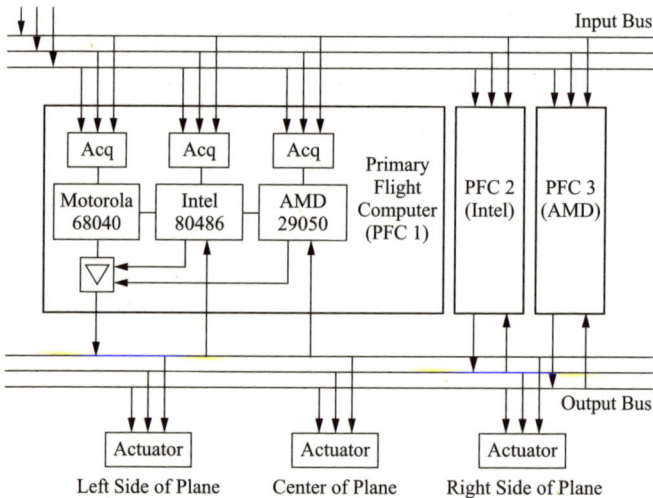

**Figure 5.26.** *Architecture of a Boeing 777*

Let us recall a few characteristics:

– each of the three lines has a different role: commanding, putting on standby and monitoring;

– the commanding line additionally has the task of selecting the outputs;

– the outputs are selected by choosing an average value;

– to achieve synchronous acquisition over the three PFCs, a synchronizing mechanism is required;

– with critical values, the results must be consolidated;

– 99% of the software is built in Ada ([ANS 83]) and three different compilers are used.

So far, we have talked about active redundancy; where all the treatment units perform their assigned treatments. Some types of architecture, however, use passive redundancy. With passive redundancy (see Figure 5.27), the passive unit is generally inactive but can be assigned auxiliary functions.

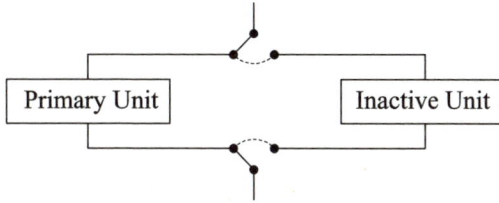

**Figure 5.27.** *Passive redundancy*

Active redundancy (W) and passive redundancy (SB) can be combined so as to increase the availability of the system (see Figure 5.28). If one of the treatment units experiences a failure, it is switched off and a reserve unit is introduced into the loop. This type of architecture is often used for systems which are difficult to access.

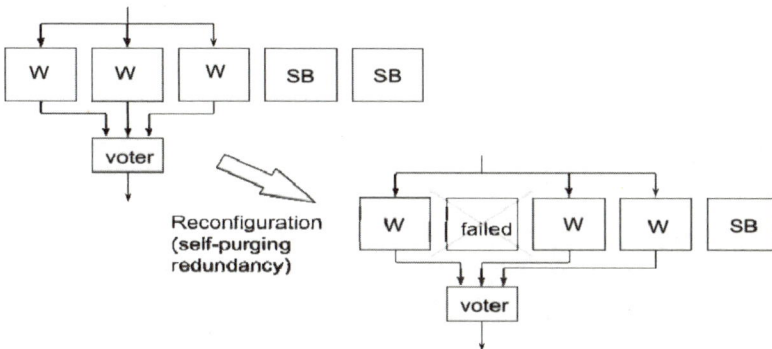

**Figure 5.28.** *Passive and active redundancy*

The use of active and passive redundancy may be dynamic. For instance as was the case with the space shuttles, unexpected events can be responded to by quickly changing the mission, as shown on Figure 5.29. The five treatment units related to the flying functions can easily be reconfigured.

In the ascending phase (a), four units are configured in the ascent mode, and one of the units is saved. As to the orbiting phase (b), two treatment units are configured to manage the flight and one unit manages the whole system. Another unit is inactive and configured with the descent management application, whereas the last unit remains as a backup. Finally, the descent phase (c) is similar to the ascent phase (a).

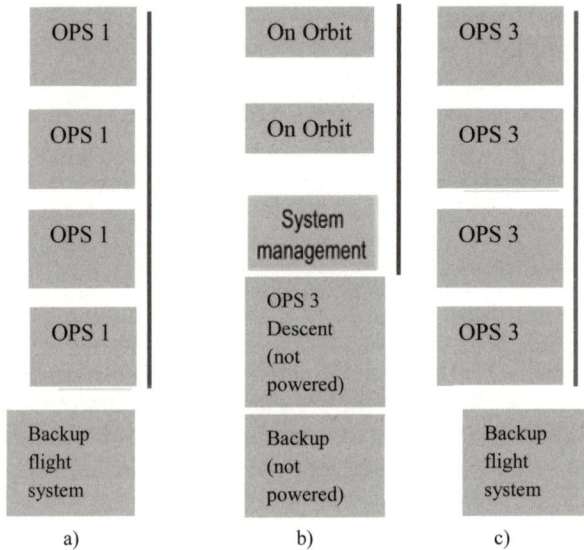

| OPS 1 | On Orbit | OPS 3 |
| OPS 1 | On Orbit | OPS 3 |
| OPS 1 | System management | OPS 3 |
| OPS 1 | OPS 3 Descent (not powered) | OPS 3 |
| Backup flight system | Backup (not powered) | Backup flight system |
| a) | b) | c) |

**Figure 5.29.** *Architecture reconfiguring*

### 5.4.3.3.5. Asymmetrical architecture

The previous section allowed us to explain *NooK* types of architecture, as well as their major flaw, i.e. cost (usage and maintenance costs). However, as shown by [DUF 05], the automotive sector is subject to different constraints, just as important: weight, congestion, power consumption, etc.

This is why, on the basis of the worker/checker types of architecture (section 5.4.3.3.3), various modes of optimization have been proposed. The first idea was expressed as follows: it is possible to attribute different objectives to each treatment unit. The "functional" unit performs its function F, whilst the second unit ("check") controls the actions of the first.

More precisely (see Figure 5.30), unit F not only executes its functions, but also performs all the self-tests. The results of these self-tests are then transmitted to a unit, $f$, which decides on the functional state of the first unit. According to the final architecture, this unit $f$ can (a) request a reset, (b) switch off the outputs, and (c) indicate that there is malfunction. As to the state of unit $f$, its good functioning is ensured through specific self-tests. These self-tests are reduced to the necessary functions.

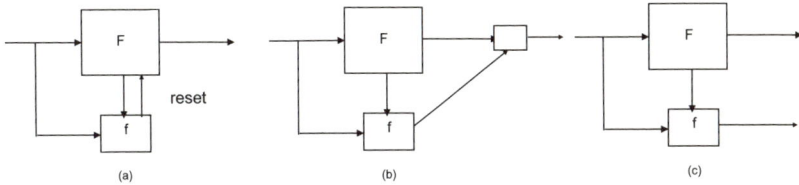

**Figure 5.30.** *Functional/check*

This type of architecture imposes at least two constraints:

– the self-tests should be chosen carefully so as to effectively detect problems in unit F;

– the software executed by unit F is categorized as "safe" and requires to be treated with care.

This type of architecture has the main advantage of reducing congestion, weight, power consumption and costs (as unit $f$ is cheaper).

The automotive sector (see chapter 9 in [BOU 11]) goes even further as it uses an asymmetrical architecture type with three-layered software.

Figure 5.31 shows an asymmetrical type of architecture with two different-sized treatment units connected to each other.

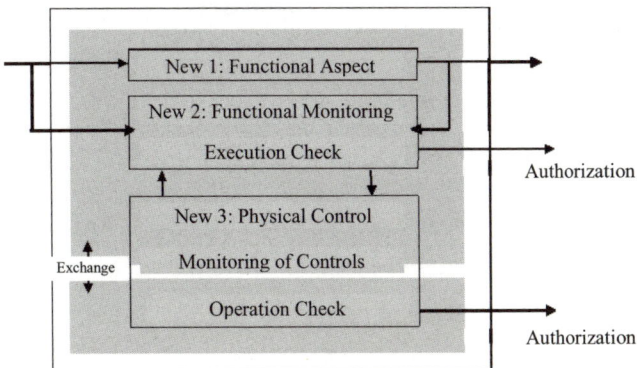

**Figure 5.31.** *Asymmetrical architecture*

The software is made up of three layers:

– level 1: deals with functional aspects by acquiring, performing the functions and generating outputs;

– level 2: this level has the responsibility of checking the functional treatment. Its objectives are to detect treatment errors from level 1 and to check what the responses are to error detection;

– level 3: this level checks that there is no malfunction in the hardware tools. This level is independent from level 2; and broken down into its two treatment units, it performs a verification based on a set of questions/answers.

Level 2 is generally based on the double acquisition of inputs, functional redundancy, verification of temporal constraints and data checks. As to level 3, the set of questions/answers is available within a ROM, and the main unit gives some incorrect answers so as to check that the control unit is able to detect errors. The main unit disposes of several trials.

This type of architecture offers several advantages:

– it proposes a clear separation between the functional aspect and safety features;

– it proposes a decomposition that can be extended to a whole sector (e.g. automotive), and can be generalized as far as a professional standard (as much in terms of software as hardware);

– it proposes a way of managing costs (financial, energy, congestion) related to the hardware tools;

– etc.

5.4.3.3.6. Summary

This section allowed us to present a considerable set of redundant architecture types. There are many different possible combinations, but some architecture types are becoming more classic.

Here is a concise list of architecture types that use redundancy (NooK or N out-of K):

– 1oo1: simplex system;

– 1oo2: duplicated system, one unit is enough to perform the function;

– 2oo2: duplicated system, all the units must be operational (fail-safe);

– 1oo2D: duplicated system (fail-operational);

– 2oo3: two units must be functional (masking);

– *NooK*: architecture with any N and K.

### 5.4.4. *Retrieval through error or error recovery*

A *NooK* type of architecture can compensate for errors. If the level of redundancy is sufficient ($N > 2$), the erroneous state of a unit does not have any impact on the system outputs. In these cases, care should be exercised as to automatic and systematic compensation (error masking), which can cloud the visibility of errors; so, signaling flaws is essential.

Error compensation can be obtained through a recovery mechanism:

– error recovery with retrieval through error: the state of the system is continually saved and allows us to return whenever errors are found;

– error recovery with tracking: if there is an error, we look for the next acceptable state, which is usually a degraded state. Tracking can only be achieved if we have a list of possible errors to address.

#### 5.4.4.1. *Retrieval through Error*

The retrieval through error mechanism consists of restoring the system to a safe previous state. In order to do so, the system must be saved regularly, and the saved systems should allow reloading. When an erroneous situation is detected, it is possible to reload one of the previous situations and to launch the execution again. If the error originates in the environment or in a transient failure, the system should be able to adopt a correct operating mode from there. If the failure is systematic (hardware or software), on the other hand, the system will go back to an erroneous mode. Some systems have several alternatives for software applications, and activate a different replica of the application when they are retrieved through errors.

The main advantage of this strategy is the possibility to delete the erroneous state, without impacting on the search for the place of error, or its cause. Following through errors can thus be used to recover unexpected errors, including design errors.

Retrieval can be either of high or low grain. The recovery by high grain restores the last overall correct condition and reruns the same software application completely. Recovery by low grain is a finer recovery strategy with the restoration of a local correct state (confined to a function, for example), alternative execution, etc.

High-grained recovery has the advantage of setting up a single point of recovery and remains suitable for systems subjected only to hardware failures.

### 5.4.4.2. *Continuation*

Continuation consists in following the execution of an application from the erroneous state, making selective corrections to the state of the system. These include making the controlled environment "safe", as it may have been damaged by the failure.

So, continuation is an activity specific to each system and depends on the accuracy of the predictions as to places and causes of errors (i.e. identifying the damage).

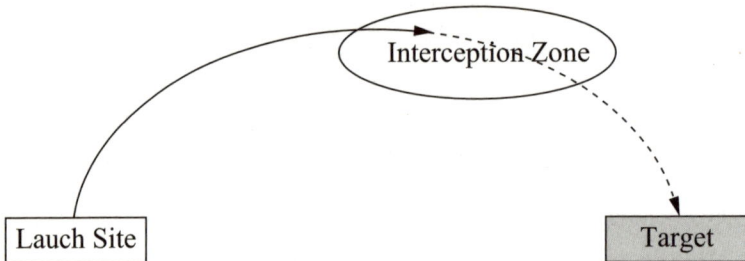

**Figure 5.32.** *Example of retrieval through error*

### 5.4.4.3. *Summary*

Retrieval through error is usually already used in practice (saving the data in banking systems, etc.). It can apply exclusively to hardware or software.

Figure 5.32 shows the example of a missile that has to go through a "busy" (wind, temperature, EM fields, explosions, radiations, etc.) interception zone before reaching its target.

In order to ensure the success of its mission, the system repeatedly saves in case of a serious flaw. It reboots and reloads its last situation. There can be a loss in precision on the trajectory which has no effect on the objective.

### 5.4.5 *Partitioning*

Partitioning is a safety technique highly recommended by the automotive sector. The current version (non-finalized) of the ISO 26262 standard [ISO 11] explains the basics of this technique (see Figure 5.33).

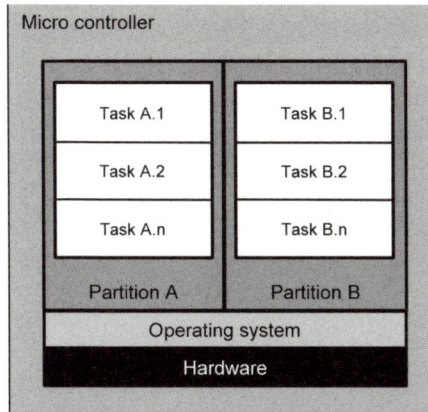

**Figure 5.33.** *Partitioning within a treatment unit*

Software partitioning must be specified in the equipment-level manufacturing phase (so as to implement protection mechanisms) and/or during the software manufacturing.

Resource sharing can only be done if it can be demonstrated that there will not be any interference among the various partitions. Either there are no interactions among the partitions and protection methods must be used to ensure impermeability among the tasks; or there are interactions and the exchanges must be shown to conform to their respective levels of safety (besides proving their impermeability).

For applications with high levels of safety associated with them (SIL 3 – SIL 4), partitioning can be achieved through dedicated architecture. It is to be noted that "software" elements in charge of partitioning management will have the same levels of safety as the partitioned software.

## 5.5. Conclusion

For hardware types of architecture, there are techniques that allow detection and even correction of errors produced both randomly and systematically. These techniques are very common in various fields (rail transport, air transport, power, process control, automotive, etc.).

This chapter has enabled us to present these techniques with relevant examples, although the high number of possibilities does not allow comprehensive enumeration. This is why we have chosen to focus on the implementation of these techniques, with the help of concrete case studies, which will be developed in the following chapters.

In this chapter, we presented some techniques used for the safety management of hardware architecture. The safety management needs to apply these processes and in each domain, there exists a standard that describes the global approach necessary, see [IEC 08] for E/E/EP application, [DO 00] for avionic, [CEN 00, CEN 03, CEN 10] for railway and [ISO 11] for automotive.

## 5.6. Bibliography

[ABR 96] ABRIAL J.R., *The B Book – Assigning Programs to Meanings*, Cambridge University Press, Cambridge, MA, 1996.

[ANS 83] ANSI, Standard ANSI/MIL-STD-1815A-1983, Langagé de programmation Ada, 1983.

[BAL 03] BALEANI M., FERRARI A., MANGERUCA L., PERI M., PEZZINI S., "Fault-Tolerant Platforms for Automotive Safety Critical Applications", *Proceedings of the 2003 International Conference on Compilers, Architecture and Synthesis for Embedded Systems*, pp. 170-177, 2003.

[BIE 99] BIED-CHARRETON D., "Concepts de mise en sécurité des architectures informatiques", *Recherche Transports Sécurité*, no. 64, pp. 21-36, July – September, 1999.

[BOU 10] BOULANGER J.-L., *Safety of Computer Architectures*, ISTE Ltd, London and John Wiley & Sons, New York, 2010.

[BOU 11] BOULANGER J.-L., *Sécurisation des architectures informatiques industrielles*, Hermes-Lavoisier, Paris, France, 2011.

[CEN 00] CENELEC, NF EN 50126 Railway applications – the specification and demonstration of reliability, availability, maintainability and safety, January 2000.

[CEN 03] CENELEC, NF EN 50129 European standard. Railway applications – communication, signalling and processing systems – safety-related communication in transmission systems, 2003.

[CEN 10] CENELEC, EN 50159-1 European standard. Railway applications – communication, signalling and processing systems, September 2010.

[DO 00] Design assurance guidance for airborne electronic hardware, ARINC, no. DO254, EUROCAE, no. ED80, 2000.

[DUF 05] DUFOUR J.L., "Automotive Safety Concepts: 10-9/h for less than 100E a piece", *Automation, Assistance and Embedded Real Time Platforms for Transportation*, AAET, Braunschweig, Germany, 16-17 February 2005.

[ESS 00] ESSAME D., ARLAT J., POWELL D., Tolérance aux fautes dans les systèmes critiques, LAAS report, no. 151, March 2000.

[FOR 89] FORIN P., "Vital coded microprocessor principles and application for various transit systems", *IFAC - Control, Computers, Communications in Transportation*, pp. 137–142, 1989.

[GEO 90] GEORGES J.P., "Principes et fonctionnement du Système d'Aide à la Conduite, à l'Exploitation et à la Maintenance (SACEM). Application à la ligne A du RER", *Revue Générale des Chemins de Fer*, no. 6, June 1990.

[GUI 90] GUIHOT G. and HENNEBERT C., "SACEM software validation", *ICSE*, PP 186–191, March 26-30, 1990.

[HEN 94] HENNEBERT C., "Transports ferroviaires: Le SACEM et ses dérivés", *ARAGO 15, Informatique tolérante aux fautes*, Masson, Paris, 1994.

[IEC 08] IEC, IEC 61508: Sécurité fonctionnelle des systèmes électriques électroniques programmables relatifs à la sécurité, international standard, 2008.

[ISO 11] ISO, ISO/CD-26262, Road vehicles – Functional safety, 2011.

[LAP 92] LAPRIE J.C., AVIZIENIS A., KOPETZ H. (DIR.), "Dependability: Basic Concepts and Terminology", *Dependable Computing and Fault-Tolerant System*, vol. 5, Springer, New York, NY, 1992.

[LEA 05] LEAPHART E.G., CZERNY B.J., D'AMBROSIO J.G., DENLINGER C.L., LITTLEJOHN D., Survey of Software Failsafe Techniques for Safety-Critical Automotive Applications, Delphi Corporation, SAE, reference 2005-01-0779, 2005.

[LIN 99] VAN LINT J.H., *Introduction to Coding Theory*, Springer, Berlin, 1999.

[MAR 90] MARTIN J., GALIVEL C., "Le processeur codé : un nouveau concept appliqué à la sécurité des systèmes de transport", *Revue Générale des Chemins de Fer*, June, 1990.

[MAT 98] MATRA and RATP, *Naissance d'un Métro. Sur la nouvelle ligne 14, les rames METEOR entrent en scène*. PARIS découvre son premier métro automatique. Numéro 1076 -Hors-Série. La vie du Rail & des transports, October, 1998.

[VIL 88] VILLEMEUR A., *Sûreté de fonctionnement des systèmes industriels*, Eyrolles, Paris, 1988.

[YEH 96] YEH Y.C., "Triple-Triple Redundant 777 Primary Flight Computer", *IEEE Aerospace Applications Conference*, 1996.

## 5.7. Glossary

| ALU: | Arithmetic and Logic Unit |
|---|---|
| COTS: | Commercial Off-The-Shelf |
| CRC: | Cyclic Redundancy Check |
| DSP: | Digital Signal Processor |
| ECU: | Electronic Computing Units |

FPU:          Floating Point Unit

IEC[7]:        International Electrotechnical Commission

MAC:          Multiply and ACcumulate

nOOm:         n Out Of m

RER:          *Réseau Express Régional* (suburban express train)

SACEM:        *Système d'aide à la conduite, à l'exploitation et à la maintenance*
(System to help production, use and maintenance)

SIL:          Safety Integrity Level

THR:          Tolerable Hazard Rate

---

7 To learn more, visit: http://www.iec.ch.

# 6

# Principles of Software Safety

The safety of an electronic control unit depends upon the hardware architecture but also upon the safety of the software. For the software, the safety management can be done by using some techniques such as redundancy, diversity, etc. But the use of this kind of technique complexifies the software and it is thus preferable to manage the design assurance level based on quality management.

## 6.1. Introduction

In Chapter 5, we presented the techniques for making aspects of hardware equipment safe, so as to meet the identified safety objectives (see Chapter 4). The next step is to make software meet those same safety objectives. It is to be noted that in several sectors, there are specific standards that target software application development: DO178C, EN 50128:2011 and IEC 60880.

To start with, we shall present the techniques to make software safe (software redundancy, data redundancy, defensive programming, error detection, forward continuation, backward retrieval, etc.) and we shall explain how they make the software more complex. From this we will be able to present the notion of quality management and its links with SSIL (software safety integrity level).

## 6.2. Techniques to make software application safe

A number of techniques can be used during the development cycle in order to manage the incorrect states of software applications. These techniques control the state of the software application and, for some they enable it to move back to a correct operating mode.

When it comes to making software applications safe, the following techniques may be used:

– error management (section 6.2.1);

– error recovery (section 6.2.2);

– defensive programming (section 6.2.3);

– double execution of the software application (section 6.2.4);

– data redundancy (section 6.2.5).

This section will outline these various techniques and, identify their strong and weak points.

## 6.2.1. *Error management*

### 6.2.1.1. *Principles*

When specifications are set up, the following challenge arises, namely "how to indicate that errors or exceptional circumstances are encountered when executing a service (function, procedure, segment of code, etc.)".

Indeed, in normal circumstances, the result of a given operation is of a certain type (for the implementation of a certain type); whereas, when exceptional circumstances arise, this result will not be any of those values because of an erroneous environment. The error cannot usually appear in a field of a returned type. Therefore, two types are incompatible.

There are several approaches that can be used to tackle this problem. The first solution consists in extending the type and thus defining a special constant, e.g. "indefinite" and, in the cases of anomalies, returning this result as "indefinite". The second solution introduces an error identification parameter, so that the operation returns an n-uplet for which one of the components indicates whether the operation was carried out correctly or not.

The first solution is of course the most elegant one, but it is only applicable for the specification level in question; as for the implementation level, it is not always possible to find an equivalent in terms of programming languages. This is why the second solution is used for the implementation (see Figure 6.1).

Identifying errors must now be associated with processing. If a given operation has not generated any errors, the error indicator is stationed at "success". In the opposite case, the value "failure" will appear. The requesting procedure must then test the error indicator so as to determine whether the request has been successful.

In successful cases, for each operation, the implementation works as seen in Figure 6.1.

```
BEGIN
...
operation(x,y,z, failure);
IF failure
    THEN – process anomaly
    ...
ENDIF
...

END
```

**Figure 6.1.** *Mechanisms of error management: function with parameter*

This kind of implementation (see Figure 6.2) may be simplified by replacing the procedure by a function that returns a status indicating success or failure of the operation.

```
BEGIN
...
IF operation(x,y,z) = failure
    THEN – process anomaly
    ...
FINSI
...

END
```

**Figure 6.2.** *Mechanisms of error management: function return*

In many cases, the processing of the anomaly ends up interrupting the execution flow and then has to branch off or return to the caller by propagating the anomaly again, as shown on Figure 6.3.

```
BEGIN
...
IF operation(x,y,z) = failure
    THEN return FAILURE
    ...
ENDIF
...

END
```

**Figure 6.3.** *Failure propagation*

Another approach to error processing is the exception mechanism through the use of modern programming languages (Ada [ANS 83], Eiffel, C++, etc.).

```
BEGIN
        ...
        operation(x,y,z)
        ...
EXCEPTION
        failure: -- process anomaly

END
```

**Figure 6.4.** *Implementation of an exception mechanism*

Managing exceptions (Figure 6.4) gives a clear coding but delocalizes error processing and increases the complexity. This complexity is produced by the attempt to generalize the exceptions, as it becomes more difficult to establish a link between a given exception and the service or the call for service that generated the exception.

### 6.2.1.2. *Summary*

Error processing is a tricky part of the design, as conventional error processing can easily be overlooked by programmers (omission of returning codes, non-processing of the error codes, systematic propagation of error codes, loss of exceptions, etc.).

Because specific treatments are added (parameters, conditions, connections, etc.) for error management, the complete handling of errors through usual methods considerably complicates the logic of the program.

Cyclomatic complexity (noted V(g)) is multiplied by at least two for each error processing. It therefore becomes very difficult to understand, maintain and test a software application with this type of technique.

### 6.2.2. *Error recovery*

### 6.2.2.1. *Presentation*

Error recovery aims to put the system back into a correct state after the detection of an error. After the detection of an error, there are two possibilities, either return (recovery through retrieval) to a correct state so as to execute an alternative of the software application, or to execute a series of actions to correct the current state and reach the right state (recovery through continuation).

Error recovery thus helps compensate for errors. There are two recovery mechanisms:

– error recovery through retrieval: the state of the system is periodically saved and can be returned to when errors have been found;

– error recovery through continuation: as an error occurs, a new acceptable state, although generally degraded, is searched for. Tracking can only be achieved if a list of errors is available.

### 6.2.2.2. *Error recovery through retrieval*

Retrieval (backward error recovery) involves restoring the system to a safe state. In order to do so, the system must be saved regularly, and the saved systems should allow reloading. When an erroneous situation is detected, it is possible to reload one of the previous situations and to launch the execution again. If the error originates in the environment or in a transient failure, the system should be able to adopt a correct operating mode from there. If the failure is systematic (hardware or software), on the other hand, the system will return to an erroneous mode. Some systems have several alternatives for software applications, and activate a different replica of the application when they go through error reprisal.

The main advantage of this strategy is the possibility to delete the erroneous state, without impacting on the search for the place of error, or its cause. Following through errors can thus be used to recover unexpected errors, including design errors.

Retrieval may be *coarse-grained* or *fine-grained*. *Coarse-grained* retrieval reinstates the last overall correct states and re-executes the same software applications from scratch. *Fine-grained* retrieval aims to adopt a finer strategy by only targeting a local state (e.g. a function), alternative execution, etc.

Coarse-grained retrieval offers the advantage that it only has one point from which to start the retrieval, and it is also adapted to systems which are occasionally subject to hardware failures.

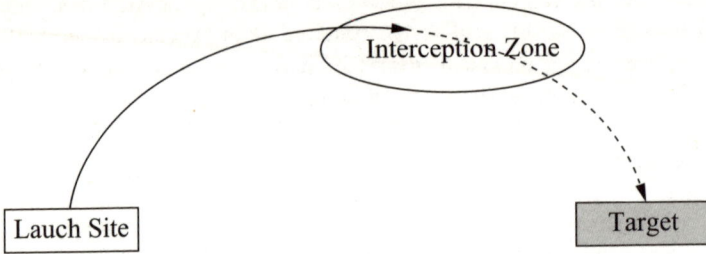

**Figure 6.5.** *Example of a system that uses retrieval*

Figure 6.5 shows the example of a missile that has to go through a "busy" (wind, temperature, EM fields, explosions, radiations, etc.) interception zone before reaching its target. The system saves its states regularly, so that, in case of unexpected reboot, the previous state can be restored so as to resume the mission. For this type of system, it is not required to have alternatives, and coarse-grained retrieval is enough.

However, in the case of systems based on error detection, the notion of alternatives is important in order to inhibit errors. A programming structure that takes into account the possible alternatives is required. The notion of an alternative brings into the frame a larger range of failures, including software design failures.

```
ENSURE <acceptance test >
BY
          <primary module>
ELSE
          <alternative module>
ELSE
          <alternative module>
          ...
ELSE
          <alternative module>
ELSE
          error
END
```

**Figure 6.6.** *Possible syntax of the recovery blocks*

Figure 6.6 is an example of syntax that describes the recovery blocks. The structure is based on identifying an acceptance test along with a series of alternatives, that are introduced in a specific order.

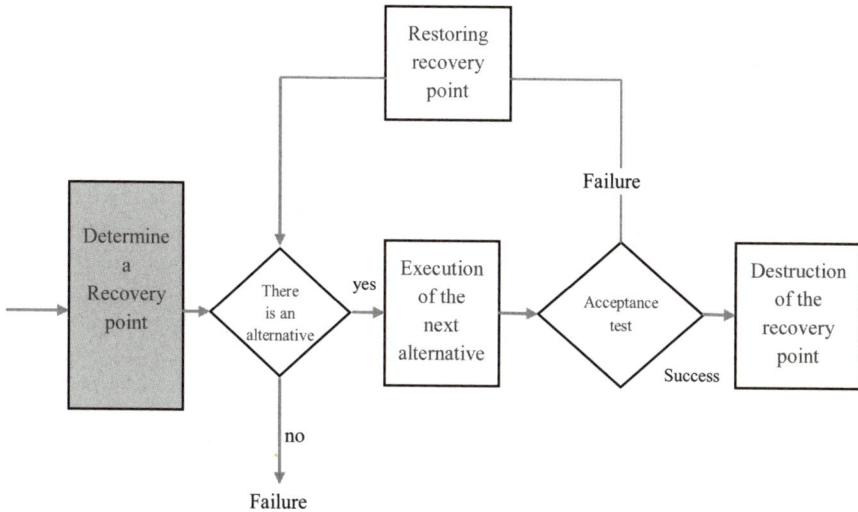

**Figure 6.7.** *Operation principles of recovery blocks*

As shown in Figure 6.7, a block starts with a point of automatic recovery that saves the current state, and ends with an acceptance test. The acceptance test (assertion) is used to check whether the system's state is acceptable after the execution of one of the blocks.

The principles for the operation of recovery blocks (Figure 6.7) are as follows:

– if the acceptance test fails, the program is restored to the recovery point at the start of the block, and an alternative module is executed;

– if the alternative module also fails the acceptance test, the program is restored to the recovery point and a different module is executed, and so on;

– if all the modules fail, the block fails as well and the recovery must start from a higher level.

```
ENSURE Rounding_err_has_acceptable_tolerance
BY
      Explicit Kutta Method
ELSE
      Implicit Kutta Method
ELSE
    error
END
```

**Figure 6.8.** *Example of a recovery block*

In the example shown in Figure 6.8, the *explicit Kutta* method is fast although inaccurate when the equations are said to be "stiff". This is why it is evaluated as an first alternative. The *implicit Kutta* method is more expensive but it can also handle "stiff" equations. The proposed solution will then be able to process all the equations and can potentially tolerate design errors in the *explicit Kutta* method if the acceptance test is flexible enough.

As shown in the second example of Figure 6.9, the recovery blocks may be embedded. If all the alternatives in an embedded recovery block fail the acceptance test, the recovery point for the external level will be restored and an alternative module to this block executed.

```
ENSURE rounding_err_has_acceptable_tolerance
BY
    ENSURE sensible_value
    BY
          Explicit Kutta Method
    ELSE
          Predictor-Corrector K-step Method
    ELSE error
    END
ELSE
    ENSURE sensible_value
    BY
          Implicit Kutta Method
    ELSE
          Variable Order K-Step Method
    ELSE error
    FIN
  ELSE erreur
END
```

**Figure 6.9.** *Example of embedded recovery blocks*

The acceptance test provides a mechanism for error detection which can then use software application redundancy. The design part of these tests is crucial to the effectiveness of the "recovery block" procedure. There needs to be a compromise between providing comprehensive acceptance tests and keeping the load associated with them to a minimum, so that flawless executions are not affected by it. It is to be noted that the term used is *acceptance* and not *accuracy;* this accounts for components delivering a degraded service.

All the techniques for error detection may be used to shape an acceptance test. Figure 6.10 shows some examples of acceptance tests. Nevertheless, these must be given the appropriate weight as a faulty acceptance test may lead to non-detection of residual errors.

---

Domain check  (assertion):
    0<= speed <= 500
Structural check:
    The last element of peak list has a "null" value
Control flow check:
    Update flag
    Launching processing
    Check of the state of the flag
Temporal check:
    Software watchdog
Information check:
                CRC, parity bit, …
Inverse check:
    Calculation of yy=squaredroot (xx) and check that yy*yy=xx

---

**Figure 6.10.** *Examples of acceptance tests*

The example in Figures 6.8 and 6.9 is a textbook example which is used in several books and articles on software safety. However, in the end, it is not representative of the problem we are trying to address.

If we have several alternatives for a given segment of code (function, procedure, program, etc.), this solution may be chosen, as there is no particular difficulty apart from an increase in maintenance costs (several versions of the bit of code), an increase in the development costs (several ways of developing and demonstrating diversity) and an increase in the complexity and testing effort.

### 6.2.2.3. *Error recovery through continuation*

Continuation (forward error recovery) implies closely following the execution of a given application from an erroneous state by making selective corrections to the state of the system. This practice includes making the checked environment "safe", as it may have been damaged by the failure.

Tracking is then an activity specific to each system and depends on how accurate the predictions are as to the position and cause of the error (i.e. damage recording).

Modern programming languages (Ada [ANS 83], Eiffel, C++, etc.) propose a mechanism referred to as exception processing (Figure 6.4), which helps managing the recovery through tracking.

Recovery through tracking is a relatively difficult technique to put into place. The principle must be used in a systematic (recovery of all exceptions) and controlled way. The challenge lies in the test: how can we demonstrate the effectiveness of the mechanism?

Let us recall that one of the reasons for the Arianne 5 rocket accident was related to the loss of exceptions, which resulted in the fall-back of two safety calculators and the subsequent destruction of the rocket.

### 6.2.2.4. *Summary*

Retrieval is widespread in practice (saving the data in banking systems, etc.). It can apply exclusively to hardware or software. Retrieval must be accompanied by at least one alternative, which makes complexity levels jump from one to two or even three along with the acceptance test and the procedure of saving and restoration of the correct state. Retrieval is relatively simple to implement and requires maintenance work on the different alternatives.

Recovery through continuation requires a list of errors and the possibility to correct these. Adding error detection points and operation check points will add to the complexity of the code and raise the combinations of tests to be performed in order to validate the application. Recovery through continuation will also make the maintenance of the software application more difficult, as each development may potentially introduce new errors and new operation check points.

### 6.2.3. *Defensive programming*

6.2.3.1. *Presentation*

In the defensive programming approach, the programmer assumes that there can be undetected errors or inconsistencies in the code. So what is defensive programming?

In line with the CEI/IEC 60880 standard [IEC 06] – section B.3a, we would add that "plausibility checks should be performed (defensive programming)". The CEI/IEC 61508 standard [IEC 00] – section C.2.5[1] is a bit more explicit on defensive programming. The aim is to "Produce programs capable of detecting command flows or erroneous data or the values of erroneous data during the execution, and to react to these errors in a predetermined and acceptable manner." and the means are: data checks (type, domain, limit, plausibility checks, etc.), plausibility checks for the inputs and intermediary variables, control of the outputs' effects through observation, integrity control of the software and execution tools.

6.2.3.2. *Principles*

Defensive programming consists of adding bits of code which check the state of the system after each modification and ensures that the change in state is consistent.

If an inconsistency is detected, several strategies may be envisaged:

– the change of state is cancelled;

– return to a correct state of the system;

– the error is diagnosed and the treatment process stopped.

Two types of defensive programming have been identified:

– *weak* defensive programming: the operation is still carried out;

– *strong* defensive programming: the software application stops.

Defensive programming can be said to be the opposite of offensive programming. The latter is based on the idea that the responsibility for checking the operating conditions of a given service goes to the person requesting that service.

Defensive programming can be implemented through four different techniques:

– establishing consistency/coherence checks for the state of the system (controls on the inputs and outputs, etc.);

---

1 This point is made almost word for word in the CENELEC EN 50128 standard [CEN 01] – section B.15.

– reporting errors (section 6.2.1);

– recovering errors through tracking (section 6.2.2.2);

– using assertion.

6.2.3.2.1. Checking consistency/coherence of state

Maintaining a given state of the system involves management of the initial state. The first method uses restrictive initialization of all the variables (general and local). The choice of values for the initialization should be related to the notion of the safe state.

Managing consistency/coherence of the software application state implies managing inputs, so as to avoid that an erroneous input propagates through the software application. This is why the second approach systematically checks the inputs (parameters of a function enquiry, general variables, function enquiry returns, etc.).

```
Root function (in xx: integer, out yy : integer): boolean
begin
            if (xx < 0) then return False
            …
            …
            return True
end
```

**Figure 6.11.** *Example of defensive programming[2]*

The "root" processing of Figure 6.11 applies the principles of defensive programming. If parameter xx is negative, the anomaly is identified and the responsibility for decision-making is delegated to the originator as to what measures should be taken in the event of failure. This typically happens with libraries, as they cannot take any decision without knowing, *a priori*, the operating environment.

```
IF (xx>0) THEN … ENDIF
IF (xx>0) THEN … ELSE No processing ENDIF
```

**Figure 6.12.** *Example of IF structure in defensive programming*

2 The language used in Figure 2.16 is the PASCAL.

The third approach is related to the management of data during the processing. In order to do this, for each programming structure, the context of each different use must be taken into account. For a IF instruction (Figure 6.12), there will systematically be a THEN branch. For a CASE instruction, there must systematically be an ERROR (see Figure 6.13).

```
CASE xx OF
1 : BEGIN ... END;
2 : BEGIN ... END;
ELSE ERROR
END
```

**Figure 6.13.** *Example of a CASE structure in defensive programming*

There are several benefits in using this type of rule. The first is the control it gives over random, as well as systematic anomalies through modification of the data. As in the development phase, a person can modify this type of *AccountType* and add a new value. This modification may be related to forgetfulness and failure to update all the structures that manipulate this type.

| enum AccountType<br><br>{<br>Savings,<br>Checking,<br>MoneyMarket<br>} | switch (accountType)<br>{<br>case AccountType.Checking:<br>    // DO something ...<br>case AccountType.Savings:<br>    // DO something ...<br>case AccountType.MoneyMarket:<br>    // DO something ...<br>default:<br>    Debug.Fail("Invalidaccounttype.");<br>} |
|---|---|

**Figure 6.14.** *Example of CASE structure in defensive programming*[3]

Defensive programming can be implemented through assertion, as shown in Figure 6.15.

---

3 Program in C#

```
int f (int x)
{
assertion (x>0) ; // precondition
...
y = ...
...
assertion (y>0) ; // postcondition
return (y) ;
}
```

**Figure 6.15.** *Example of assertion in C*

6.2.3.2.2. Plausibility of the data

Do the values of the variables appear plausible given the knowledge we have of the program? Plausibility is interesting as it allows us to avoid going through the whole process again to check the validity of the data. We are trying to verify the properties. It is to be noted that the acceptance tests in Figure 6.10 are similar to plausibility tests.

When it comes to the coherence of input data, it is based on the fact that there is redundancy among the inputs as a whole:

– two distinct acquisitions, for example by the same chain or by two distinct chains;

– a given acquisition is unique but with different sources for the final information, e.g. speed is measured through two wheels with different gears;

– a piece of information may be coded;

– inputs may be contradictory;

– etc.

For some systems, there are post-conditions (see e.g. Figure 6.15) which help checking the coherence of the treatment. These post-conditions make the link between the outputs and the inputs.

Some of these post-conditions address particular aspects (e.g. measuring an angle) and are related to issues of physics (maximum acceleration), to implementation choices (the last element of a list is always the null element), etc.

For example when measuring the angle of a steering wheel in a car, this angle cannot physically change by more than 90° over the cycle.

6.2.3.2.3. Plausibility of operation

Techniques related to plausibility of operation revolve around the question "Does the execution of the program follow a predictable and expected flow?" The techniques applied are:

– the analysis and verification of signatures (as much as for the code as for the data);

– error detection.

Coherence in the operation helps assessing whether a given software application follows a previously validated path. In order to do so, the execution of the software application must be traceable. The execution is composed of a number of *traces,* each trace being a series of checkpoints (execution path).

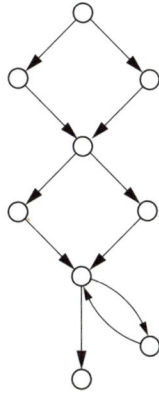

**Figure 6.16.** *Program structure*

The first graph of Figure 6.16 represents a program with two IF instructions after a WHILE instruction. Analysis of the operating execution paths can lead to the selection of two execution traces. These traces are characterized by the creation of checkpoints. These indicators can be local (several pieces of information are stored within them) or general (only one variable is handled).

As an example (see Figure 6.17), it is possible to have a variable starting at 0, which, when it switches to a THEN branch, is increased by 1; when it switches to ELSE it is decreased by 1; and which, by the end of the execution must be equal to 2 or -2.

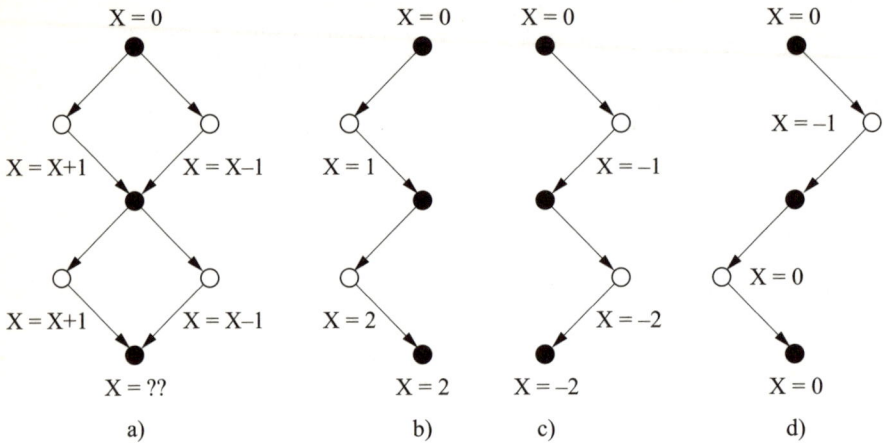

**Figure 6.17.** *Execution trace check*

In some systems with high levels of criticality, the checkpoints are in high numbers and help in controlling the final stage of the operation, or even the entire operation.

The more numerous and/or complex these traces are, the higher the number of checkpoints. This has consequences on the memory space (storage of the traces and of the current trace), on the execution time (complementary treatments), as well as on the complexity of the code (addition of processing unrelated to functional aspects, makes the code and the analyses associated with it more complex).

### 6.2.3.2.4. Assertion management

An assertion is a condition which must be observed in order for the program to be pursued. Assertions may serve as pre-conditions or post-conditions.

A pre-condition (Figure 6.18) ensures that the execution of a software application is interrupted if some conditions are not respected. Pre-conditions are above the procedure (functions and/or methods). It is the most common method for defensive programming. The basic idea is that "there is no execution if nominal conditions are not met".

```
Public void print(String message)
{
assert message!= null: "Error: message is null";
System.out.println(message);
}
```

**Figure 6.18.** *Example of pre-condition in C*

Using assertions as post-conditions (Figure 6.19) is much rarer. Post-conditions ensure that the contract is respected from the point of view of the client. They are often positioned at the end of the procedure (functions and/or methods). For instance, the client is ensured that the method will never give him a blank list.

```
Public List<String> filter(List<String> mails)
{
...
assert myList!= null:"Error:null list";
return myList;
}
```

**Figure 6.19.** *Example of post-condition in C*

To summarize, assertions bring an additional load to the execution of a software application. This load results from the verification of unexpected events. It is to be noted that some languages enable inhibition of assertions, so that all protections are removed.

### 6.2.3.3. *Summary*

Defensive programming is one of the techniques required for almost all the standards and is widely used. However, both defensive programming and flaw management are sometimes combined. Indeed, when identifying situations where the software application does not respond correctly, it is possible to use palliative codes without understanding the anomaly fully.

Defensive programming can be exposed to two problems:

– identifying problematic situations (dividing by zero, taking the squared root of a negative number, etc.) for the whole or part of the software application;

– identifying actions to correct these while avoiding systematically sending the problems back to the requesting services.

The first consequence of defensive programming is the much higher operation time. Due to its nature, defensive programming, as it enhances the complexity of the code, also makes testing it and maintenance more difficult. As far as testing is concerned, defensive programming generates situations which are:

– either difficult to test: some flaws are taken into account by several barriers so that it is impossible to trigger internal barriers (protected by other barriers);

– or impossible to produce: the sequence of failures is an example of a difficult test to apply.

### 6.2.4. *Double execution of the software application*

#### 6.2.4.1. *Principles*

Operational redundancy uses twice the same application through the same treatment unit (processor, etc.). The results are usually compared through a device external to the processor and any inconsistency provokes a fall-back of the arithmetic unit (*fail stop* behavior). This technique is often used for a programmable logic controller (PLC).

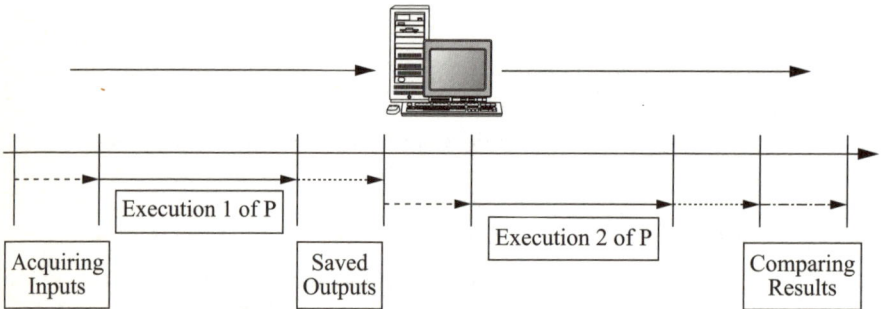

**Figure 6.20.** *Principles of operational redundancy*

Figure 6.20 shows a timeline for operational redundancy. The triplet "acquisition, execution, saving" appears twice and, in the end, a comparison of the saved results is made.

The first instance of operational redundancy can help detecting memory failures. A unique program is loaded in to two different zones of the storage medium (two different zones in the addressing memory, two different memory supports, etc.).

Memory failures (RAM, ROM, EPROM, etc.) can then be detected along with intermittent failures of treatment units.

It is to be noted that some failures of the shared hardware devices (comparison unit, treatment unit) are not detected and therefore stay latent. There are two ways in which errors can be masked:

– the outcome of the comparison of results can always be positive, no matter what the inputs are (failure of the means of comparison is a common mode of failure);

– subtle failures (e.g. A-B is systematically performed instead of A+B) in the ALU part in the treatment unit can give the same erroneous result for each operation (the treatment unit is a common mode of failure).

```
int F(a,b)                          int F(b,a)
{                                   {
    int y = 0;                          int y = 0;
    y = 2*a+b+1                         y = -(-2*a-b-1)
    return y                            return y
}                                   }
```

Figure 6.21. *Diversification of the code*

One solution is to introduce self-tests during the operation; e.g. a comparison of data inconsistency (usually without the need to enter a fall-back state), comprehensive functional tests for the treatment unit. If we want effective error detection, the coverage of the tests must be wide enough (cover the usage instructions, etc.) and the tests must be performed at the appropriate moment (initialization, with each cycle, regularly, at the end of a mission, etc.). The major drawback with this solution, however, is its cost in terms of performance (related to the space taken by the self-tests as well as their frequency).

Another solution consists of introducing diversification of the code. This diversification can be "light" and we refer to it as voluntary dissymmetry of the encoding application. It is then possible to enforce the use of two different sets of instructions to program the application. For instance, one of the programs would use A+B, whereas the other would use -(-A-B) as shown in Figure 6.21. A dissymmetry in the data can be introduced by addressing the stored objects differently for each of the two programs (variables, constants, parameters, functions and procedures).

| xx | yy |
|---|---|
| a+b | - (a-b) |
| A or B | not (not(A) and not(B)) |
| .... | |

**Table 6.1.** *Example of diversification rule*

It is to be noted that this voluntary dissymmetry can be introduced automatically within a single program. For these two types of dissymmetry the compilation phase must be controlled as well as the final executable to make sure that the dissymmetry is still present.

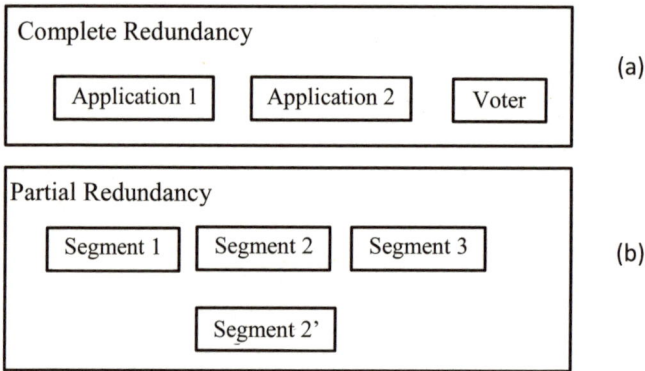

**Figure 6.22.** *Complete redundancy (a) and partial redundancy (b)*

Generally speaking, complete redundancy (where the whole application is executed twice) is used; however, partial redundancy may be enough (Figure 6.22(b)).

Operations using floating variables within a safety function require the use of specific safety techniques. Partial redundancy relies on diversification, which can, for example, use different libraries. We must then be tolerant to errors in the comparison.

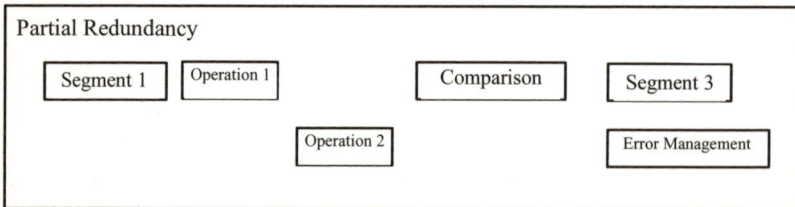

**Figure 6.23.** *Partial redundancy*

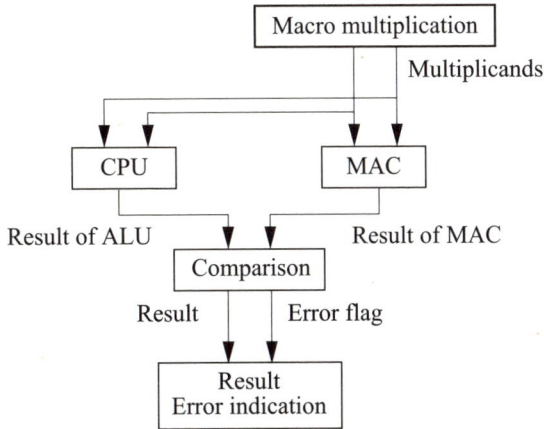

**Figure 6.24.** *Process of double execution and comparison*

Software diversification may be complemented by hardware diversification. As shown in Figure 6.23, a type of hardware architecture (see Figure 6.24) can use the FPU[4] of the main processor with an additional unit to execute operations. The code is then slightly more diversified as two sets of instructions are used instead of one. The notion of tolerance to error then becomes crucial.

### 6.2.4.2. *Example 1*

Introducing partial dissymmetry is relatively simple but has only a weak error detection power. It is possible to generalize this solution through diversification of the application in question. There are various possibilities, e.g. using two teams, or two different code generators, etc.

As an example, Figure 6.25 shows how the EBICAB 900 works. The application being used is diversified into applications A and B. Application A is divided into F1, F2, F3; whilst application B is divided into F1', F2', F3'. Three development teams are then necessary; two independent teams are in charge of creating the two applications, whereas a third team is assigned the responsibility of the specification (which is common to the two) and of synchronization. As there is only one acquisition phase, the data are protected (CRC[5], etc.). The data manipulated by

---

4 FPU stands for floating point unit.
5 Cyclic redundancy check ( CRC) is an easy and powerful way of controlling the integrity of the data.

application A are diversified (different memory, mirror bit-by-bit, etc.) compared to application B.

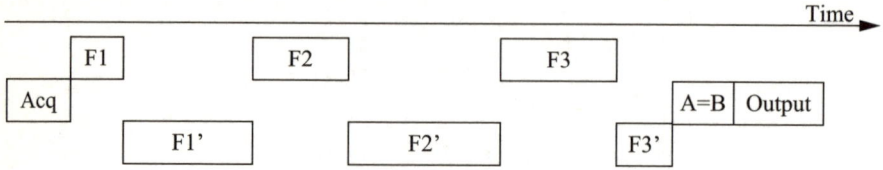

**Figure 6.25.** *Principles of temporal execution of the EBICAB*

### 6.2.4.3. *Example 2*

As our second example, we shall draw on a piece of field equipment which relates the central computer to the markers (elements giving commands to trains). In this application, only one software application is formally developed with method B [ABR 96] and a treatment unit.

Method B is a formal method which ensures (through mathematical evidence) that the software is correct in terms of property. This warranty is interesting but does not cover the code generator, the chain generating the executable (compiler, linker, etc.) or the loading means.

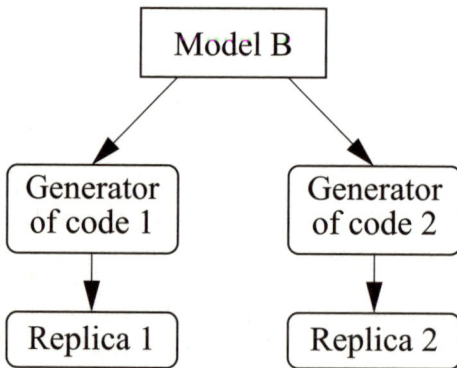

**Figure 6.26.** *Diversification*

As shown in Figure 6.26, for this application there are two code generators and two interpreters executing the code (two compilers). This gives two different versions of the code and address tables (variables, constants, functions, parameters, etc.) which can be shown to be actually distinct for both executables. Each version of the application is loaded in different memory spaces.

### 6.2.4.4. *Example 3*

In this example we shall turn to information consoles onboard trains. In the case of consoles displaying speed-related information, the feared event is "displaying erroneous control orders or speed information".

One possibility to ensure safety (SIL-2) of the display is to use reactive safety [CEN 03]. This technique imposes the condition that the detection and passivation time should not exceed the specified limits for a potentially dangerous transitory output.

Let us recall the definition of reactive safety, in the words used in the CENELEC EN 50129 standard [CEN 03]:

– "This technique enables safety-related functionality to be performed by a simple entity, on the condition that its functioning is safe by rapid detection and passivation of any dangerous breakdown. Although the entity performs the effective safety-related functionality, the detection/check/test functionality should be considered as a second, independent entity, so as to avoid any common breakdown."

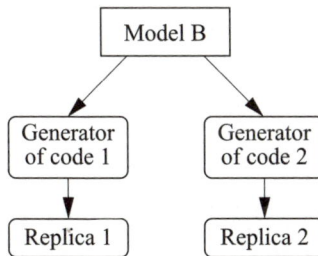

**Figure 6.27.** *Diversification of supporting libraries*

The maximum tolerated duration for erroneously permissive information must be lower than or equal to the limit of perception of the human eye.

A console is made up of a treatment unit which holds two applications. Each of these builds up the image to be displayed. As a general rule, an image is composed of critical and non-critical information.

A comparison mechanism helps in checking that the two applications have produced a similar image. The term "similar" is an indication that the comparison is being performed on "critical" parts. According to the power available, we can either compare the whole image or only the minimum required. The comparator has become a common mode of failure, just as the voter can be in this type of architecture.

As to the comparison, it can be performed among produced images, or between one of the produced images and the display memory (VRAM), so that display memory failure can be detected.

The main challenge with graphical application is that building an image requires going through a library which is a COTS. Graphical libraries are difficult to validate and contain a large number of flaws.

In fact there is only one application of the SIL-2 level but it is linked to two different libraries which offer graphic elements based on different algorithms.

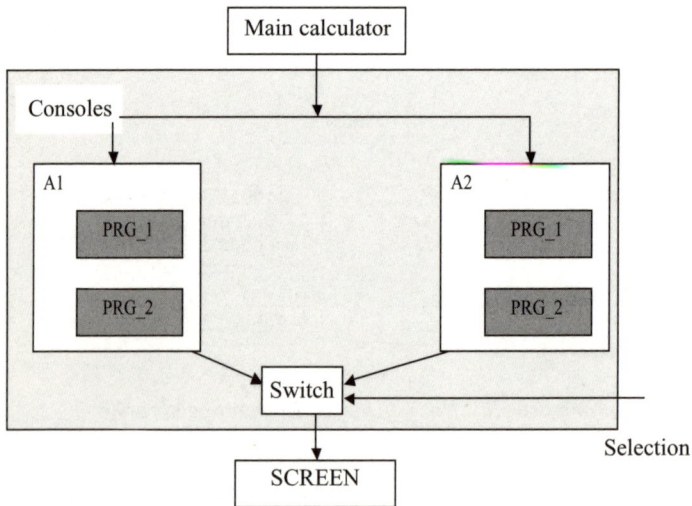

**Figure 6.28.** *Safe and available architecture*

Diversification of the supporting libraries and operation redundancy can be combined with a unit redundancy so as to relate safety to a high level of availability, as shown on Figure 6.28.

As can be seen in Figure 6.28, the system is made up of two identical consoles named A1 and A2. One of these is said to be the "main" console, whereas the other one is referred to as the "secondary" console. In the event of the main console failing, a switch inactivates the second one.

### 6.2.4.5. *Example 4*

An independent element of safety, sometimes referred to as the safety bag, is a component that intercepts actions requested by users or a component of the system, and checks their validity according to the safety rules established during the development phase (it is a sort of plausibility check). The independent component of safety has its own representation of the state of the system. This representation should be precise and recent enough to be able to apply the whole set of rules it oversees. This technique is touched upon in [KLE 91].

### 6.2.4.6. *Summary*

Operational redundancy is a fairly simple technique which has the advantage of using a single treatment unit. Its main disadvantage on the other hand remains the long operating time it requires, as launching the double processing with vote and self-test requires at least 2.5 to 3.5 times the time of a single processing. This is why this type of solution is primarily used for systems whose processing time is not an issue.

Partial or total diversification of the code allows a good error detection rate for both random and systematic errors (depending on the degree of diversification), although the cost will increase (as there will be two software packages to maintain, etc.)

### 6.2.5. *Data redundancy*

#### 6.2.5.1. *Presentation*

We have explained in section 6.2.2 the notion of error recovery, and in section 6.2.4, that of double execution. Both these techniques may imply redundancy of the software application. Data redundancy, on the other hand, only focuses on the data handled by the software application.

Data redundancy works by identifying the data that needs protection. Either all or only a subgroup of data will need to be protected.

In its simplest form, data redundancy duplicates the stored variables and compares their content. Once the list of data to be saved has been identified, the next step is to assign this data twice. The duplication can be managed through a safety mechanism which makes the process transparent. In order to do so, the writing and reading services that will manipulate the two images must be defined. Then, the check points which will analyze the consistency between the two images are chosen.

```
F(...) :...
START
...
WRITE (A, 10)
...
WRITE (B, READ (A)+READ(B))
...
END
```

**Figure 6.29.** *Example of a program using data redundancy*

Figure 6.29 shows an example of function *F*, which writes the value 10 in variable A, the value *A+B* in variable B. In order to do so, the services read and write are used. Figure 6.30 shows an example where these same services are applied to manage the two replicas of one variable.

```
WRITE (x, v)
START
    V1(x) = v
    V2(x) = v
END

READ (x)
START
IF V1(x) = V2(x)
            THEN RETURN (V1(x))
            ELSE ERROR
ENDIF
END
```

**Figure 6.30.** *Implementation of the writing operation*

When it comes to the two images, they may be identical or different. If the replicas are identical, a slippage test should be used. The replicas may complement each other (mirrors). In that case, the sum of the two replicas should be null $(V1(x) + V2(x) = 0)$.

Diversification of the replicas allows inclusion of random and systematic failures of the hardware tools (processor, memory,etc.). Indeed, a slippage test with zero as a reference is easier to validate than the same test between two random values.

In order to ensure the innocuousness of a memory failure on the data, it is important to diversify the zones of the memory, its banks and even the type of memory.

### 6.2.5.2. Generalization

In [GOL 06 – chapter 2], a more systematic mechanism of data redundancy is applied through a general algorithm (or a tool) to a whole bit of code and/or software application.

Three rules should be applied:

– any variable $x$ is duplicated into $x0$ and $x1$;

– any writing operation is performed on $x0$ and $x1$;

– after each reading of variable $x$, both copies should be checked.

| BEGIN | BEGIN |
|---|---|
| A := B; | A0 := B0; |
| END | A1 := B1; |
| | IF (B0 != B1) |
| | THEN ERROR(); |
| | ENDIF |
| | END |

**Figure 6.31.** *Protection of a simple assignment*

Figures 6.31 and 6.32 shows the results of the use of the above algorithm on two very simple programs which can be summarized in a single assignment.

| | |
|---|---|
| BEGIN<br>   A := B+C;<br>END | BEGIN<br>   A0 := B0+C0;<br>   A1 := B1+C1;<br>   IF (B0 != B1) OR (C0 !=C1)<br>      THEN ERROR();<br>   ENDIF<br>END |

**Figure 6.32.** *Protection of a double assignment*

Generally speaking, this algorithm is relatively simple to apply (Figures 2.29 and 2.30). However, when using some languages such as language C, the management of the functions and/or procedures may lead to more profound changes, as shown in the example of Figure 6.33.

| | |
|---|---|
| F(A :INTEGER) :B :INTEGER<br>BEGIN<br>  B := A+1<br>END | F(A0,A1 :INTEGER,<br>B0,B1 :INTEGER)<br>BEGIN<br>   B0 := A0;<br>   B1 := A1;<br>   SI (A0 != A1)<br>      THEN ERROR();<br>   ENDIF<br>END |
| VARIABLE<br>  X,Y : INTEGER;<br>BEGIN<br>  X := 1 ;<br>  Y := F(X) ;<br>END | VARIABLE<br>X0,X1,Y0,Y1: INTEGER;<br>BEGIN<br>  X0 := 1 ;<br>  X1 := 1 ;<br>  F(X0,X1,Y0,Y1) ;<br>END |

**Figure 6.33.** *Protection of a function*

The interesting thing about this generalization is that it can be applied through an application that takes as an input a software application and gives as an output a software application with duplicated data.

### 6.2.5.3. *Summary*

Duplicating data is relatively simple to do but it makes the code more complex. The use of a library or automatic duplication ensures the maintainability of the code and the management of complexity.

It is to be noted that duplicating each variable can be avoided by introducing a protection mechanism of a CRC-type, which is a form of information redundancy.

## 6.3. Other forms of diversification

In the previous sections, we have introduced the notions of redundancy and diversification. Diversification helps in the increasing of effectiveness in terms of the rate of failure detection, and in avoiding a common mode. Common modes of failure may lead to non-detection of failure because different parts of the system are affected in an identical or similar way by the failure.

Diversity can thus become an element used to demonstrate that the safety objectives for a given piece of equipment are met. On top of hardware diversification (which gives heterogeneous redundancy) and diversification of the code, it is also possible to put in place temporal and assignment diversification in memory.

### 6.3.1. *Temporal diversity*

Temporal diversity involves building a redundant type of hardware architecture so that the various operation units do not execute the same code at time t.

The need for diversity can be associated with different types of failures:

– systematic failure at the level of the operation unit and processing failure;

– etc.

### 6.3.2. *Diversity in memory assignment*

Diversity in memory assignment consists of building a hardware type of architecture in such a way that the elements stored in the memory (local variables, general variables, constants, application parameters, binary, parameters of exchange between functions, etc.) are never allocated to the same spaces.

The requirement for diversity is linked to different types of failure:

– when using a common operating system such as DOS and DOS extender (to manage a larger memory space), it may be necessary to diversify the memory allocations so as to avoid common modes as regards management of extended memory;

– etc.

## 6.4. Overall summary

The maintainability of a software application is, after its safety, the most important task. Indeed, a software application has a more or less extended lifetime according to the sector (15 years for the automotive sector, 40 to 50 years for railway and aircraft industries, 50 years for the nuclear industry,etc.). The ability to maintain a software application is related to safety and is hence essential.

From this discussion, we gather that the only activity that can ensure the safety and maintainability of a software application is quality management. Quality management is achieved through quality checks and safety checks. A dependable software application is thus a simple application. It is crucial to limit complexity in a software application to a strict minimum.

This is why all the standards (DO 178 [ARI 92, ARI 11], CEI 61508 [IEC 08], CENELEC EN 50128 [CEN 01, CEN 11], ISO 26262 ([ISO 11]), IEC 60880 [IEC 06], …) concerning software applications with safety-related objectives, introduce the notion of the *safety level* and advocate quality management as a basic technique, through the ISO 9001:2000 standard [ISO 00], along with *pre-established* and *systematic* techniques.

Chapter 3 allowed us to analyze various standards and to define the notion of the safety level. Quality management will be the focus of the next section.

## 6.5. Quality management

### 6.5.1. *Introduction*

Although quality is supposed to *help,* it is often considered by the members of a project as:

– a "word", it does not serve any purpose, it is not visible in the project, it does not add anything, …

– a plague, it is a waste of time, creates more problems than would otherwise be the case, it is a straightjacket, etc.

The quality is a prescriptive or controlling activity which ought to be understood and accepted. It should provide methods, processes, and help control the activities. Quality management makes *pre-established* and *systematic* activities available for use.

Through *pre-established* and *systematic* aspects of quality standards, competence and efficiency emerge. Competence is obtained by applying and understanding the treatments; whereas efficiency is obtained by proposing improvements, understanding and accepting the difficulties met, and establishing lessons-learned processes.

### 6.5.2. *Completing the software application*

It is to be noted that we use the word *completion* of a software application and not *development.* Completion includes developing but also checking, validating, producing, installing, and maintaining the software application.

Checks and validation are important and will be more or less complex according to the level of safety required. When it comes to the final production of the application and its implementation, these are crucial activities and require following specific procedures. Removing a software application is mentioned but does not pose any problems, as opposed to removing complex systems, such as a nuclear power plant or a railway installation.

Maintaining the software application remains the most delicate activity as it develops, a certain level of safety should be maintained while managing the costs of this development and minimizing the impact on the system in operation.

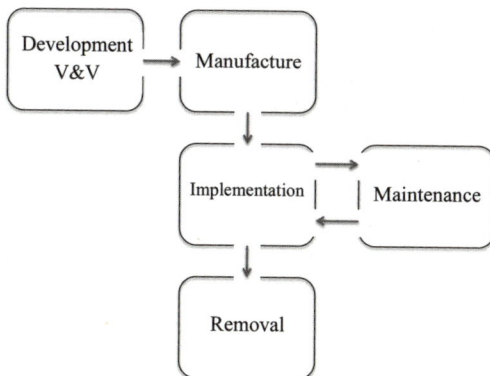

**Figure 6.34.** *Completion of a software application*

Maintenance in a software application can be challenging because of the duration of its lifetime. For the railway sector, the lifetime will be 40 to 50 years; for aerospace, 40 years, for the nuclear industry 50 years, and finally for the automotive sector, 15 years. Given the differences in the duration of life for these sectors, measures should be taken to ensure maintenance of the service and of the software application in question.

### 6.5.3. *Completion cycle*

As previously mentioned, completing a software application requires several steps (specification, design, coding, testing, etc.). We refer to the lifecycle of the software application. This lifecycle is useful to describe the dependencies and chains of activities. The lifecycle should include progressive refinement of the development as well as possible iterations. In the following section, we shall present a lifecycle used to complete a certified software application.

6.5.3.1. *V-model and other completion cycles*

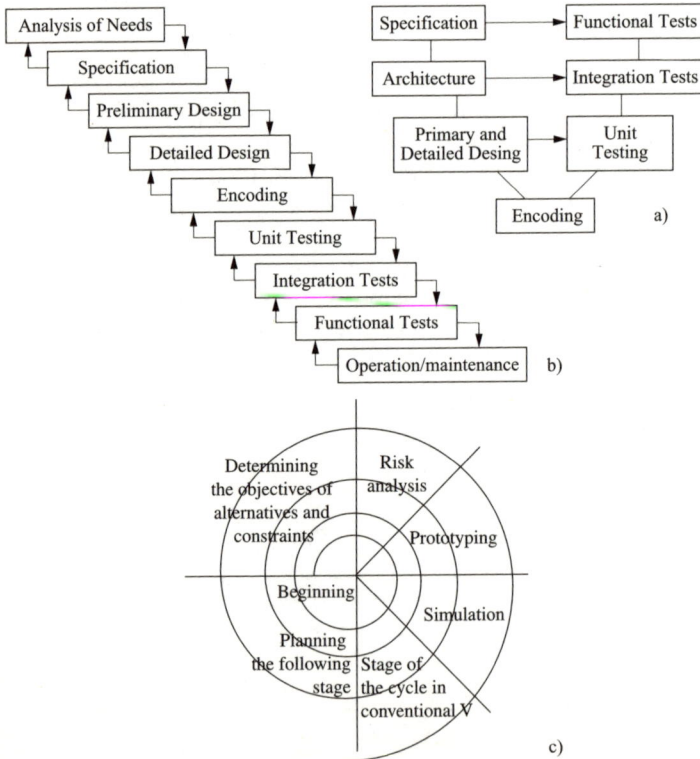

**Figure 6.35.** *Three lifecycles*

As can be gathered from Figure 6.35, there are several cycles ((a) V-model, (b) cascade cycles, (c) spiral cycles, etc.) for the completion of a given software application. However, the most recommended cycle (EN 50128 [CEN 01a, CEN 11], DO 178 [ARI 92, ARI 11], IEC 61508 [IEC 08], ISO 26262 [ISO 11]) remains the V-model.

Figure 6.36 shows a V-model such as it is generally presented. The analysis of the needs aims to check that the expectations of the client are met within technological feasibility. The specification phase on the other hand, aims to describe what the software should do (not how is should do it).

When defining the architecture, we try to decompose the software application hierarchically into its module/components, and we identify the interfaces between those elements. During its design, each module/component (data, algorithms, etc.) are described. Often the design phase is done in two steps. The first step, named preliminary design, identifies the required data and services. The second phase, referred to as detailed design, describes all the services through their algorithms. The design phase then leads to the encoding phase.

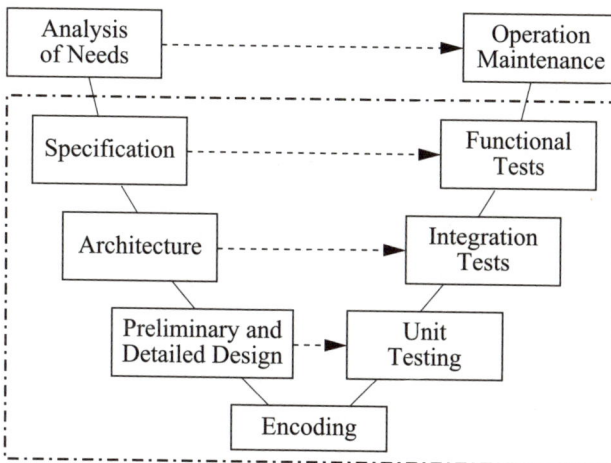

**Figure 6.36.** *V-model*

Figure 6.36 shows that there are several phases to the tests: unit testing (focused on the low-level components), integration tests (for the software and/or hardware interfaces), and finally functional tests. These aim to show that the product conforms to its specifications. As to the operating/maintenance phase, it concerns operations and the management of possible solutions.

It is to be noted that there is a horizontal correspondence (dotted arrows) between specification activities, and activities of design and testing. The V-model is thus made up of two phases, the descending and the ascending phase. Activities of the ascending phase should be prepared during the descending phase. Figure 6.37 is thus closer to the recommended V-model.

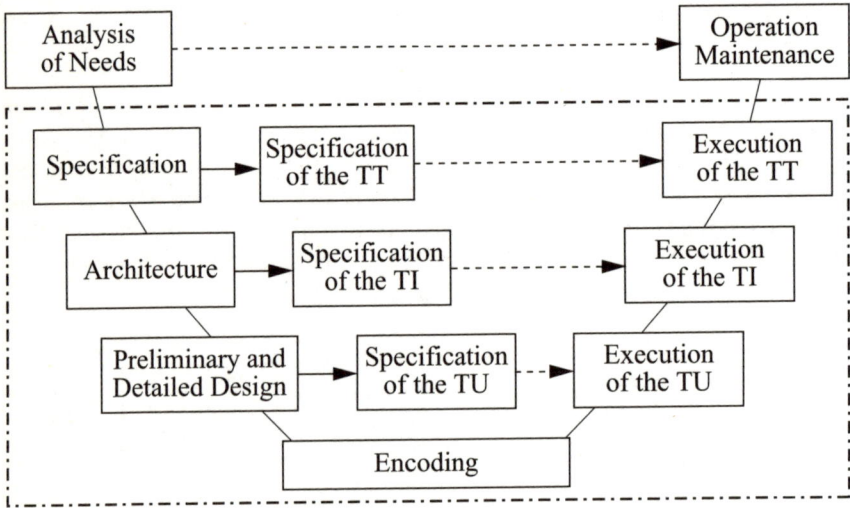

**Figure 6.37.** *V-model including specification tests*

### 6.5.3.2. *Verification and validation*

Completing a software application should include the design of the application, but also all the activities that show a certain level of quality has been achieved. This level of quality implies that no flaw has been introduced during the design phase, and that the product corresponds to the identified needs.

As shown in Figure 6.38, verification looks for flaws within the V-model, whereas validation demonstrates that the product corresponds to what it is required for, hence its place in the upper part of the V-model.

### 6.5.3.3. *Summary*

To conclude this section, quality is not just a "word" (and certainly not a "plague"), but is also a process. This process should apply to each step in the completion of the software application, and it should enable management of both the product and its components (hardware, software, etc.).

Quality is thus related to two important concepts: "pre-established" and "systematic". The process, procedures and guides should be *pre-established*, and they should be used *systematically*. Quality management in software is obtained through process management.

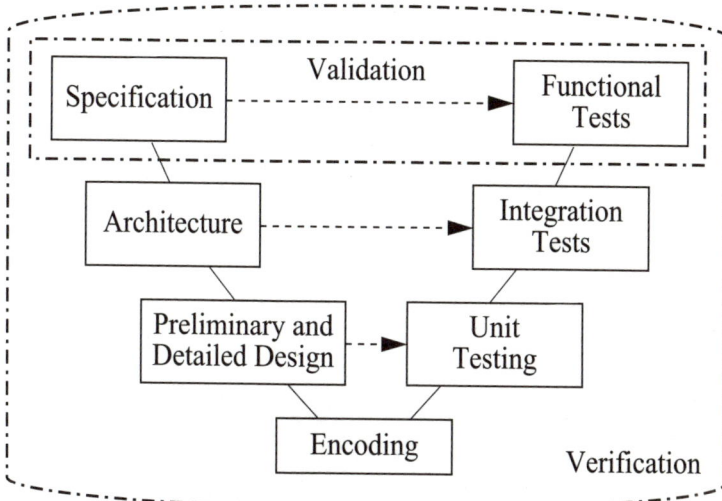

**Figure 6.38.** *Verification and validation on the V-model*

## 6.6. Conclusion

This chapter has allowed us to show the techniques required to make a software application safe. However, it appears that all of these techniques lead to an increase in the cost of development (and therefore in the length of deadlines). More importantly they involve an increase in the complexity of the software application, which in turn makes the V&V phase more complex and poses a challenge to maintenance of the software application.

For software with safety objectives (DAL, ASIL, SSIL), it is therefore preferable to put effort in to the quality management of the software. Quality management involves a development cycle and associated processes. In addition, there should be an organization to ensure a certain degree of independence among the teams, which is related to managing the competences of the people involved.

Finally, in order to obtain software with a low number of flaws, or even "flawless" software, the strictest development techniques should be implemented, such as formal techniques based on simulation of evidence and/or of model-checking (which helps in creating a safe software package, rather than trying to prove that it is not safe through testing, for example).

## 6.7. Bibliography

[ABR 96] ABRIAL JR., *"The B Book - Assigning Programs to Meanings"*, Cambridge University Press, Cambridge, MA, 1996.

[ANS 83] ANSI, Standard ANSI/MIL-STD-1815A-1983, Langage de programmation Ada, 1983.

[ARI 92] ARINC, "Software Considerations in Airborne Systems and Equipment Certification", publié par l'ARINC, no. DO 178B, et par l'EUROCAE, no. ED12, édition B, 1992.

[BAI 08] BAIER C., KATOEN J.-P., *Principles of Model Checking*, MIT Press, Cambridge, MA, 2008.

[CEN 01] CENELEC, NF EN 50128, Applications Ferroviaires. Système de signalisation, de télécommunication et de traitement – Logiciel pour système de commande et de protection ferroviaire, July 2001.

[CEN 11] CENELEC, NF EN 50128, Applications Ferroviaires. Système de signalisation, de télécommunication et de traitement – Logiciel pour système de commande et de protection ferroviaire, July 2011.

[FOU 93] FOURNIER J.-P., *Fiabilité du logiciel*, Hermes, Paris, France, 1993.

[GEF 98] GEFFROY J.-C. and MOTET G., *Sûreté de fonctionnement des systèmes informatiques*, InterEditions, Masson-Dunod, 1998.

[GOL 06] GOLOUBEVA O., REBAUDENGO M., SONZA REORDA M. and VIOLANTE M., *Software Implemented Hardware Fault Tolerance*, Springer, 2006.

[IEC 00] IEC, IEC 61508: Sécurité fonctionnelle des systèmes électriques électroniques programmables relatifs à la sécurité, international standard, 2000.

[IEC 06] IEC, IEC 60880: Centrales nucléaires de puissance – Instrumentation et contrôle-commande importants pour la sûreté – Aspects logiciels des systèmes programmés réalisant des fonctions de catégorie A, 2006.

[IEC 08] IEC, IEC 61508: Sécurité fonctionnelle des systèmes électriques électroniques programmables relatifs à la sécurité, international standard, 2008.

[ISO 08] ISO, ISO 9001:2008, Systèmes de management de la qualité – Exigence, December 2008.

[ISO 11] ISO, ISO/CD-26262, Road vehicles – Functional safety, 2011.

[KLE 91]  KLEIN P., "The safety-bag expert system in the electronic railway interlocking system elektra", *Expert Systems with Applications*, 3(4): 499–506, 1991.

[LAP 92] LAPRIE J.C., AVIZIENIS A., KOPETZ H. (eds), *Dependability: Basic Concepts and Terminology, Dependable Computing and Fault-Tolerant System*, vol. 5, Springer, New York, 1992.

[LEV 95] LEVESON N.G., *Safeware: System Safety and Computers*, Addison Wesley, Nenlo Park, CA, USA, 1$^{st}$ edition, 1995.

[LIS 95] LIS, Laboratoire d'Ingénierie de la Sûreté de Fonctionnement. *Guide de la sûreté de fonctionnement*, Cépaduès, 1995.

[VIL 88] VILLEMEUR A., *Sûreté de Fonctionnement des systèmes industriels*, Eyrolles, 1988.

## 6.8. Glossary

COTS:  Commercial Off-The-Shelf

CRC:  Cyclic Redundancy Check

IEC[6]:  International Electrotechnical Commission

nOOm:  n Out Of m

SACEM:  *Système d'Aide à la Conduite, à l'Exploitation et à la Maintenance* (System to help production, use and maintenance)

SIL:  Safety Integrity Level

SSI:  Software SIL

---

6 To learn more, visit: http://www.iec.ch.

# Certification

In some fields such as railways or aeronautics, certification can be done before the authorization of operation. Certification is based on an independent assessment of the product. ISO standards 45011 and 17020 defined the rules to manage a certification. A certificate is produced by an accredited body and the assessment respects certain rules. In this chapter, we introduce the notion of independent assessment and of certificates, focusing on the railways sector.

## 7.1. Introduction

In this chapter, we shall explore the certification process (with a reference to ISO standards 45011 and 17020) and the impacts this has on safety demonstration. We shall present the notion of cross-acceptance, as it will allow us to define a framework of acceptability for certificates from other sectors.

## 7.2. Independent assessment

Assessing ([BOU 07, BOU 06]) a product (system or software) consists of assessing how well that product complies with standards (generally, a standard, part or a set of standards), following a prescribed method. Independent assessment for instance is performed through an ISA (Independent Safety Assessor).

DEFINITION 7.1 (ASSESSMENT). – *Evaluating a product involves performing a conformity analysis of a product with a set of standards. This conformity analysis follows a pre-defined procedure.*

From definition 7.1, it can be seen that evaluating a product takes as inputs two elements: a set of standards and all the elements produced during the completion of the product.

The result of the evaluation should pronounce whether or not each of the requirements of the standards is met.

– If the product complies with a requirement, the assessment will pronounce it "complies with the requirement".

– If the product does not comply with the requirement, the assessment will state "deviation".

– If the product complies with each of the requirements of a standard, then the assessment will state that there is consistency with the standard.

– If one or more requirements are not met, the product cannot be declared consistent with the standard.

There can be several types of deviations, classified according to the "incurred risk", when the requirement is not met.

Classifying the "incurred risk" is a very subjective task, and is only used with standards that clearly distinguish mandatory requirements, from "optional" or "recommended" ones.

This allows the assessor to make an unquestionable classification of the deviations:

– a deviation from a mandatory requirement challenges the product consistency with the standard;

– whereas a deviation from a recommended requirement does not, in itself challenge the overall consistency with the standard.

Assessing a component gives rise to an assessment report. Certification consists of making the results of the assessment formal within a consistency certificate.

## 7.3. Certification

Certification is an administrative action which translates the confidence acquired by the appropriate body in the ability of the system to fulfill its safety functions. In the aeronautics sector, certification is realized by authorizing the use of the system.

Certification relies on the evidence of safety produced for the body in charge by the responsible correspondent, as defined on the organization document.

DEFINITION 7.2 (CERTIFICATION). – *Certification consists of producing a certificate which is a commitment as to the consistency of a product with a set of standards.*

*Certification is based on the results of an assessment and the production of a certificate.*

A certificate (see definition 7.2) helps define a usage perimeter as well as a perimeter of responsibilities. It is to be noted that without a certificate, a company can be made to demonstrate that the product can be introduced within the system it manages. In that case, we refer to homologation.

## 7.4. Certification in the rail sector

### 7.4.1. *Obligations*

The regulation framework in the rail sector has been complemented in recent years by standards (CENELEC EN 50126, EN 50128 and EN 50129, EN 50159, etc.), by European and national decrees, and finally by European bodies such as ERA (European Railway Agency[1]) and other national bodies (for example in the French railway market) such as EPSF (Public Railway Safety Establishment[2]) and the STRM-TG (Technical Service of Ropeway and Guided Transports[3]).

As highlighted in [BOU 08, BOU 09b], commissioning a rail system (whether urban, mainline, freight or high speed) requires approval. Approval may rely on certificates of conformity or independent assessment of the product(s).

Authorized commissioning of a rail system goes through the national body in charge of analyzing the safety file. For the French market:

– STRM-TG for the urban domain;

– EPSF for the rail domain.

In both cases, the safety file contains a description of the system and its use, safety elements and operation, and an opinion on the degree of safety of the system, given by an independent body named EOQA (expert or qualified accredited agency). For the French urban domain, the objectives of EOQA are defined in the note [STR 06].

To learn more about the effective implementation of this set of rules, we recommend [BOU 07].

---

1 To learn more about ERA, please visit http://www.era.europa.eu/.

2 To learn more about EPSF, please visit http://www.securite-ferroviaire.fr/.

3 To learn more about STRMTG, please visit http://www.strmtg.equipement.gouv.fr/.

### 7.4.2. *Needs*

These standards help assess how well the making of a product complies with a set of standards. This evaluation may lead to certification. Several bodies can provide this type of certificate through their authority EN 45011. The EN 45011 standard defines the general requirements related to those bodies.

As far as the European rail network and its conformity with technical specifications of interoperability (TSI) are concerned, Europe only recognizes notified bodies (at least one per country) which are accredited to the EN ISO/IEC 17025 standard [ISO 05]. Each country must have at least one notified body. In the case of France, the notified body is CERTIFER[4].

In both cases, the safety file comprises of a description of the system and its mode of use, safety elements for the operator and a notice on the level of safety of the system given by the independent body EOQA (expert or qualified accredited body). For the urban domain, the EOQA's aim is defined through note [STR 06].

To learn more about an effective application of these regulations, we recommend [BOU 07].

### 7.4.3. *Applying certification*

Certifying a given rail product P according to a set of standards S aims to analyze consistency of the product with the standards S. The set of standards S may include a technical reference (e.g. ERTMS[5] specification [DG 06]) and/or obligations related to the completion process (the CENELEC standards [CEN 00, CEN 01a, CEN 11, CEN 03]). Certification is a lengthy and costly procedure (taking from a few months to a few years).

The costs derive from the rigorous respect of the standards (activity lists, list of elements and overall management), the cost of the evaluation in itself (evaluation as in CENELEC [CEN 00, CEN 01a, CEN 11, CEN 03] consists of assessing the final product and not just the completion process) as well as the restart induced by the certification process (minor or major corrections so as to correct the identified non-compliances).

---

4 To learn more about the CERTIFER organization and its activities, we recommend http://www.certifer.asso.fr/.

5 ERTMS stands for European Rail Traffic Management System, for more information see t.http://www.ermts-online.com

The bodies in charge of certification should be independent from the provider. It is to be noted that in some countries it is acceptable for a certifying team to be part of the same company (e.g. as happens in Germany). For some products such as components of the ERTMS [DG 06], the national authorities have the responsibility of designating the bodies in charge of the certification process. In that case, they are referred to as notified bodies: NoBo.

In order to perform a certification in the rail transport sector, a set of standards compatible with EN 45011[6] ([IEC 98]) and/or the ISO/IEC 17020 standard[7] [ISO 12] according to the type of product, should be used, along with accreditation.

Accreditation is obtained through bodies such as COFRAC[8] (COmité FRançais d'Accréditation, French Committee of Accreditation). The association CERTIFER[9] is accredited according to the ISO/IEC 17020 standard by COFRAC and recognized by the state as the notified body (NoBo).

### 7.4.4. Implementation

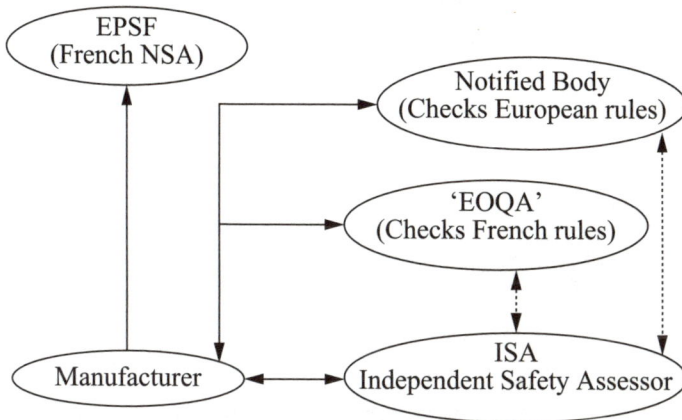

**Figure 7.1.** *Assessment, EOQA, ISA*

---

6 The EN 45011standard defines the general requirements relative to the bodies responsible for the certification of the products.

7 The ISO/IEC 17020 standard identifies the general criteria for the functioning of different types of inspecting bodies.

8 To learn more, please visit http://www.cofrac.fr/

9 To learn more, please visit http://www.certifer.fr/

As can be seen in the following figure, with new projects, the actual commissioning of the product can only be done after a safety appraisal, formalized through a safety-case. The CENELEC EN 50129 standard [CEN 03] identifies the structure of the Safety-Case.

The safety-case contains different elements including the safety assessment report. The safety evaluation must be performed by a third party, independent (EOQA) from both the requesting and providing parties.

### 7.4.5. *Upgrading management*

Upgrading management can only be performed if the software is maintainable and can be tested. These two properties have been identified as crucial by the CENELEC EN 50128 standard [CEN 01, CEN 11] but they remain difficult to apply.

One of the characteristics of rail systems lies in their lifespan. A typical duration for this type of system is 40 years with a tendency to last up to 50. The question then arises of how a software application can be maintained for 40 years.

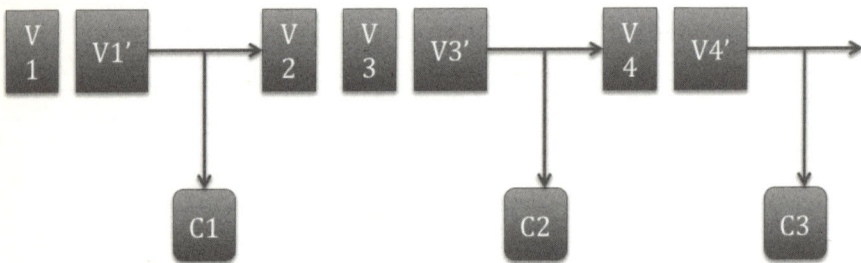

**Figure 7.2.** *Certification sequence*

Figure 7.2 shows that certification generally requires a number of restart operations, hence version Vx'. The use of delta for certification can include several upgrades (see version V2+V3).

This incremental procedure raises several issues:

– how to commercialize the products when there are several certificates?

– after several delta certifications, is it necessary to perform a comprehensive certification (if so, what are the criteria for this decision)?

– etc.

It is fairly easy to ask the above questions; however to find the answers to them, there needs to be a specific procedure to manage upgrades and their impact on the products already in use. There needs to be a set of principles to help avoide and/or manage incompatibilities (software/OS, software/software, software/hardware).

This is why the CENELEC EN 50128 standard [CEN 01a] has a whole chapter on software maintenance (chapter 16), which is usually badly understood by manufacturers, as they often consider that they are not responsible for maintaining the product during the lifetime (40 to 50 years) of the system.

Chapter 16 of the CENELEC EN 50128 standard [CEN 01a] only sets the bases for upgrade management, it does not however include software maintenance and only gives two techniques, as shown in Table 7.1. This is why in the new version [CEN 11], the maintenance chapter has been grouped with the deployment chapter.

| TECHNIQUE/MEASURE | SSILO | SSIL1 | SSIL2 | SSIL3 | SSIL4 |
|---|---|---|---|---|---|
| 1. Impact Analysis | R | HR | HR | M | M |
| 2. Data Recording and Analysis | HR | HR | HR | M | M |

**Table 7.1.** *Techniques of software maintenance*

Manufacturers managing product lines will introduce variants. This is especially true for products designed according to the legislation of a particular country and which has to be extended to another country.

Figure 7.3 shows the ideal situation (from the point of view of development, and not from the financial point of view). From a product P, characterized by a version V1 and a certificate C1, we can upgrade the product and create a variant that follows its development.

In this case, the absence of interaction between the two projects will impact on the costs (two versions developed on the same basis) and can affect safety, as the flaws detected in a given project are not necessarily the same as for the other project.

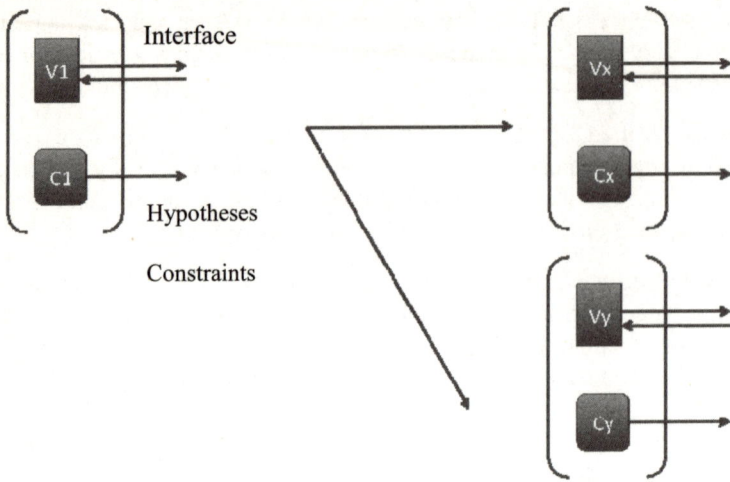

**Figure 7.3.** *Introduction of a variant*

As shown in Figure 7.4, the software application can be partitioned into two different parts: the generic part common to all the projects, and the specific part. This way, the aforementioned problem can be solved and the costs rationalized.

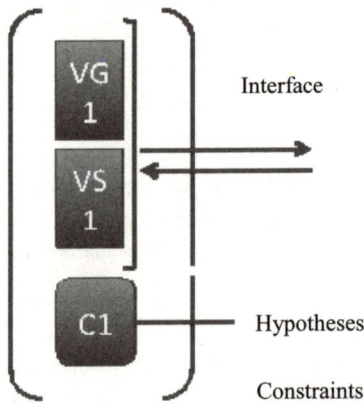

**Figure 7.4.** *Description of a specific product*

As an example, an application such as the automatic subway of the Parisian metro – line 14[10] is broken down into three pieces of equipment:

– ground equipment for local management (PAS) which only controls a section of the track. So, in order to manage the entire line, several PAS are required;

– onboard equipment (PAE) which controls the movement of a given train. There will have to be as many PAEs as there are trains;

– ground equipment for the line management (PAL), which allows operators to communicate with the whole equipment present on the line (PAS and PAE).

The three kinds of equipment interact and should be able to control the whole line. This system must be installed on other sites; this is why all application software is made up of two parts:

– a generic part which implements the services of the system;

– a specific part which describes the topology of the line.

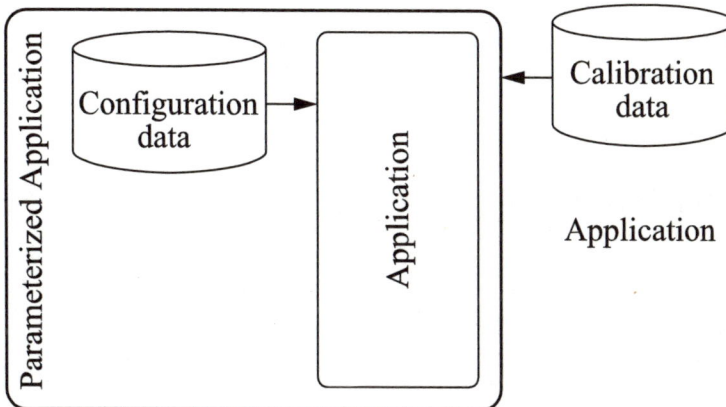

**Figure 7.5.** *Two possible types of parameterization data*

---

10 Readers interested in details on design and safety of the system SAET-METEOR for the Parisian metro line 14, please consult [BOU 06].

The CENELEC EN 50128 standard [CEN 01a, CEN 11] introduces the notion of application parameterized by the data (see the Figure 7.5), and an entire chapter has been dedicated to the principles of safety for this type of application. It is to be noted that for some applications, there might be a need for two types of parameterizing: of the configuration data and of the calibration data (which changes over time).

The calibration data are related to the fact that characteristics of the system change over time (decrease in fissionable matter for nuclear power plants, change in engine characteristics due to wearing, etc.).

Defining a generic application ([BOU 99, BOU 00]) remains an ongoing issue. Should all the possible types of behavior be included in the generic application, or only as few as possible?

If the generic application includes all the potential behavioral responses, the final software application will contain a significant number of dead codes. Generally, this code is dead through parameterizing.

The presence of dead code is prohibited or should be minimized in the case of applications with an impact on human lives. In order to show that this dead code is innocuous, the whole generic application will have to be checked and validated. Most of the time, it is not possible to thoroughly check the generic applications that contain all potential behavioral responses.

If the specific application only contains the minimum, important behavioral responses will be delegated to specific parts, and the workload for a specific project will increase. Moreover, when a new project is started, the closest specific project should be used.

The best solution consists of adopting a median position by defining an accessible generic application and for the specific applications to complement the generic one.

Whatever choice is made, the parallel development of variants will require a mechanism for the reporting of flaws from one variant to the other. A flaw regarded as dangerous for one particular variant could be trivial for another; however, an analysis should still be performed.

Figure 7.6 shows that for each stage of the system, there is a phase that checks effective compatibility of the data with what the software can take in.

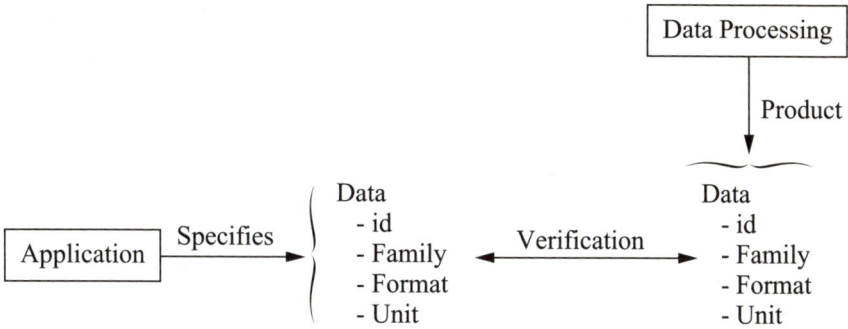

**Figure 7.6.** *Data compatibility*

### 7.4.6. *Cross-acceptance*

It should be pointed out that a certificate of conformity for a given rail product is based on a safety-case consistent with the CENELEC EN 50129 standard [CEN 03].

The structure of the safety-case for the rail sector is defined in the CENELEC EN 50129 standard [CEN 03]. The safety-case can identify hypotheses, constraints and restrictions to the use of the product. As to the certificate, it can also point out hypotheses, constraints and usage restrictions identified during the certification process.

As shown in Figure 7.7, a product is not only the description of a set of interfaces, but it also includes a number of hypotheses, constraints and restrictions.

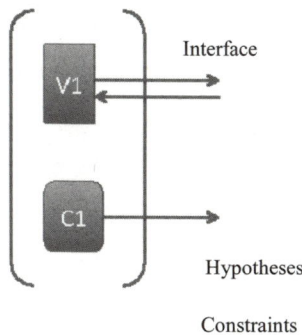

**Figure 7.7.** *Example of a certificate*

Generally speaking, designing a rail system on the basis of certified products requires compatibility checks between interfaces, between constraints, hypotheses and usage restrictions. If the standards used are not specifically made for the railway sector, the standard compatibility should also be checked.

"Cross-acceptance" consists of checking that the normative context for a certificate and the hypotheses made in it are compatible with the field of use of the product. The application guide for the CENELEC EN 50129 standard [CEN 07] describes the principles of cross-acceptance.

### 7.4.7. *Tests*

Certifying a rail product requires a certain amount of work related to design, verification, validation, and demonstration of safety and quality.

The notion of tests is an intrinsic part of the verification and validation processes.

There are different types of tests for the railway sector:
– tests of the equipment type;
– tests of the equipment series;
– tests of qualification for environmental constraints (vibration, EMC, noise, etc. See the EN 50155 standard);
– fire/smoke tests;
– tests of subsystem validation;
– tests of validation for the whole system (these can be performed on a particular test site or on the final site;
– tests of qualification for the tools and testing means; etc.

One of the constraints relative to laboratory tests is compliance with the EN ISO/IEC 17025 standard [ISO 05] which defines the general requirements for a lab authority in calibration and tests.

It is to be noted that this standard is applicable to all organizations proceeding to tests and/or calibration. For example, labs for first, second and third parties, as well as labs where tests and/or calibration are part of the product control and certification processes.

Additionally, it is to be noted that legal texts defining constraints related to authorization of commissioning highlight the need to comply with the EN ISO/IEC 17025 standard.

## 7.5. Automatic systems

The IEC 61508 standard [IEC 08] does not refer to any type of certification; however, it is common for programmable-automaton-based equipment to certify products with or without a CE (Conformité Européenne/European conformity) label.

The EC Machinery Directive (IEC 62061) requires a CE type of labelling to assure the integrity of a machine. This is a self-certifying process on the part of the manufacturer.

## 7.6. Aircraft

For civil and military aircraft, independent authorities such as the FAA (Federal Aviation Administration), and in Europe, the European Air Safety Agency is responsible for certifying the planes and their embedded systems in aspects related to safety. It is to be noted that the certification these agencies provide are not geared towards the standards EN 45011 and/or EN 17020 and that there is no mentioning of specific accreditation.

The notion of certification applies to a certain type of plane or engine. Certifying authorities treat the software as an integral part of the system or of the equipment onboard the plane or the certified engine. This means that the certifying authority does not consider the software to be an autonomous product. Currently, software and hardware are included as elements of a system.

The notion of the safety-case, which formalizes safety demonstration is absent from the aircraft industry.

## 7.7. Nuclear

Any country that produces nuclear energy has a safety authority that sets the rules to protect the population and the environment from the negative effects of radiation.

In France, it is the Nuclear Safety Authority (NSA), with technical support from the Institute for Radiation Protection and National Safety (IRPNS).

Nuclear power plants are not subject to certification, but to authorization of establishment and use. These authorizations are delivered by government decrees following propositions by the NSA.

Only government decrees must be obeyed. They formulate the high-level requirements without entering in to technical detail. In practice (in France), for systems related to computer-safety, documents published by the AIEA and the TC45 of IEC are proposed by the operator as a way of demonstrating compliance, and after considered acceptable by the IRPNS and the NSA.

## 7.8. Automotive

The automotive sector does not currently recognize the need for certification. However, the ISO 26262 standard [ISO 11] identifies a number of requirements referred to as confirmatory measurements so as to ensure that the safety assessment is carried out correctly.

The people involved in these confirmatory measurements should have a high level of minimum independence from those involved in the development and production procedures. The level of independence required increases with increasing Automotive Safety Integrity Level (ASIL) requirements.

It is to be noted that contrary to other sectors, the highest level of required independence may be reached within the same company. These measures follow three types:

– reviews checking the main products;

– an audit checking the relevance of the process;

– a safety evaluation to assess the overall safety-case.

## 7.9 Spacecraft

The standards applied in the European spacecraft sector correspond to those elaborated by the ECSS (European Cooperation for Space Standardization). It results from a joint effort by the European Space Agency, national space agencies and European space industries to develop and maintain common standards.

It is to be noted that applying the ECSS standards is not mandatory in a legal sense but is proposed and generally adopted, with occasional adaptations for each case through contractual terms.

## 7.10. Safety case

A safety-case is defined as the documented demonstration that a system and/or part of a system meets the safety (-innocuousness) objectives within a given operating environment.

The safety-case aims to be part of the demonstration that the system is dependable. It is crucial to evaluation and/or certification. The safety-case is generally produced incrementally during the different phases of the product. It may be available at some key moments of the project, such as at conception, the manufacturing, installing and/or acceptance, so as to ensure safety for each change of phase.

The safety-case should bring together all the elements that show the system complies with the standards. The safety-case will have to be updated during the whole lifetime of the system, from its design to its removal.

The safety-case is required in some sectors, such as the railway [CEN 00] and the automotive sectors [ISO 11]. It is to be noted that for the railway sector, the CENELEC EN 50129 standard [CEN 03] defines thoroughly the content of the safety-case.

According to the field and/or country, the content of the safety-case can be different. For example see [EUR 06] for aircraft control. It can be as simple as a collection of information produced during the completion of the system, or as comprehensive as possible while being completely autonomous and delivering the elements gathered in the completion process. As for aircraft, nuclear plant and spacecraft, we considered that even though a safety-case is not identified as such, the documents that justify compliance with the standards and safety requirements of the system are equivalent to it.

## 7.11. Conclusion

This chapter has allowed us to describe the issue of certification. Certification in the railway and aircraft sectors is a lengthy and relatively costly process which should be addressed as early as possible.

It is to be noted that formalized safety demonstrations are required more and more, and the notion of the safety-case is becoming generalized (see section 7.10).

Finally, and to conclude, it should be pointed out that successful examples of certification are related to the management of competences, the implementation of organizations with an appropriate level of independence, the formalizing of the procedures and a demonstration that these are respected. These procedures must cover quality management, development, verification, validation and safety management.

## 7.12. Bibliography

[BLA 08] BLAS A. and BOULANGER J.-L., "Comment Améliorer les Méthodes d'Analyse de Risques et l'Allocation des THR, SIL et Autres Objectifs de Sécurité", *LambdaMu 16, 16ème Congrès de Maîtrise des Risques et de Sûreté de Fonctionnement*, Avignon 6-10 October 2008.

[BOU 00] BOULANGER J.-L. and GALLARDO M., "Processus de validation basée sur la notion de propriété", LamnbdaMu 12, 28-30 March 2000.

[BOU 06] BOULANGER J.-L., Expression et validation des propriétés de sécurité logique et physique pour les systèmes informatiques critiques, Compiègne University of Technology, May 2006.

[BOU 06b] BOULANGERJ.-L. and SCHÖN W., "Logiciel sûr et fiable : retours d'expérience", *Revue Génie Logiciel*, no. 79, December 2006.

[BOU 07] BOULANGER J.-L. and SCHÖN W., "Assessment of Safety Railway Application", *ESREL 2007*.

[BOU 08] GALLARDO M. and BOULANGER J.-L. "Poste de manoeuvre à enclenchement informatique : démonstration de la sécurité", *CIFA, Conférence Internationale Francophone d'Automatique*, Bucharest, Romania, November 2008.

[BOU 09] BOULANGER J.-L. (ed.), *Sécurisation des architectures informatiques – exemples concrets*, Hermès-Lavoisier, Paris, France, 2009.

[BOU 09b] BOULANGER J.-L., "Le domaine ferroviaire, les produits et la certification. », *Journée « ligne produit" 15 October 2009*, Ecole des mines de Nantes.

[BOU 10] BOULANGER J.-L., "Sécurisation des systèmes mécatroniques. Partie 1", dossier BM 8070, *Revue technique de l'ingénieur*, November 2010,

[BOU 11] BOULANGER J.-L. (ed.), *Sécurisation des architectures informatiques industrielles*, Hermès-Lavoisier, Paris, France, 2011.

[BOU 11a] BOULANGER J.-L., "Sécurisation des systèmes mécatroniques. Partie 2", dossier BM 8071, *Revue technique de l'ingénieur*, April 2011.

[BOU 99] BOULANGER J.L., DELEBARRE V. and NATKIN S., "METEOR: Validation de Spécification par modèle formel", *Revue RTS*, no. 63, pp. 47-62, April-June 1999.

[CEN 00] CENELEC, NF EN 50126, Applications Ferroviaires. Spécification et démonstration de la fiabilité, de la disponibilité, de la maintenabilité et de la sécurité (FMDS), January 2000.

[CEN 01a] CENELEC, NF EN 50128, Applications Ferroviaires. Système de signalisation, de télécommunication et de traitement – Logiciel pour système de commande et de protection ferroviaire, July 2001.

[CEN 01b] CENELEC, EN 50159-1, Norme européenne, Applications aux Chemins de fer: Systèmes de signalisation, de télécommunication et de traitement – Partie 1: communication de sécurité sur des systèmes de transmission fermés, March 2001.

[CEN 01c] CENELEC, EN 50159-2, European standard, Applications aux Chemins de fer: Systèmes de signalisation, de télécommunication et de traitement – Partie 2: communication de sécurité sur des systèmes de transmission ouverts, March 2001.

[CEN 03] CENELEC, NF EN 50129, European standard, Applications ferroviaires: systèmes de signalisation, de télécommunications et de traitement systèmes électroniques de sécurité pour la signalisation, 2003.

[CEN 06] CENELEC, EN 50121, Railway applications – Electromagnetic compatibility, 2006

[CEN 07] CENELEC, European standard, Railway Applications – Communication, Signalling and Processing systems – Application Guide for EN 50129 – Part 1: cross-Acceptance, May 2007.

[CEN 07a] CENELEC, CLC/TR 50506-1 Railway applications – Communication, signaling and processing systems – Application, Guide for EN 50129 – part 1 : Cross-acceptance, May 2007.

[CEN 11] CENELEC, EN 50128, European standard, Applications Ferroviaires. Système de signalisation, de télécommunication et de traitement – Logiciel pour système de commande et de protection ferroviaire, October 2011.

[CEN 11a] CENELEC, EN 50159, European standard, Applications aux Chemins de fer: Systèmes de signalisation, de télécommunication et de traitement – Communication de sécurité sur des systèmes de transmission, August 2011.

[DG 06] DG ÉNERGIE ET TRANSPORT, Ertms - pour un trafic ferroviaire fluide et sûr, Technical report, European Commission, 2006.

[EUR 06] EUROCONTROL, Safety Case Development Manual, 13/10/2006.

[IEC 98] IEC, BS EN 45011:1998, General requirements for bodies operating product certification systems, 1998.

[IEC 01] IEC, Centrales nucléaires - Instrumentation et contrôle commande des systèmes importants pour la sûreté - Prescriptions générales pour les systèmes, IEC 61513, Edition 1.0, 22/3/2001.

[IEC 06] IEC, Centrales nucléaires de puissance - Instrumentation et contrôle-commande importants pour la sûreté - Aspects logiciels des systèmes programmés réalisant des fonctions de catégorie A, IEC 60880, Edition 2.0, 2006-05.

[IEC 09] IEC, Centrales nucléaires de puissance - Instrumentation et contrôle-commande importants pour la sûreté - Classement des fonctions d'instrumentation et de contrôle-commande, IEC 61226, Edition 3.0, 2009-07.[ISA 05] ISA, "Guide d'Interprétation et d'Application de la Norme ICE 61508 et des Normes Dérivées IEC 61511 (ISA S84.01) et IEC 62061", April 2005.

[ISO 05] ISO, EN ISO/IEC 17025 Exigences générales concernant la compétence des laboratoires d'étalonnages et d'essais, 2005.

[ISO 08] ISO, ISO 9001:2008, Systèmes de management de la qualité – Exigence, December 2000.

[ISO 09] ISO, ISO/DIS26262, Road vehicles – Functional safety – not published, 2009.

[ISO 12] ISO, ISO/IEC 17020:2012, Conformity assessment – Requirements for the operation of various types of bodies performing inspection, 2012.

[STR 06] STRMTG, Mission de l'Expert ou Organisme Qualifié Agrée (EOQA) pour l'évaluation de la sécurité des projets, version 1 du 27/03/2006.

## 7.13. Glossary

CENELEC:[11] *Comité Européen de Normalisation Électrotechnique* (European Committe for Electrotechnical Standardization)

COFRAC:    *COmité    FRançais    d'Accréditation*    (French    Accreditation Committee)

COTS:    Commercial off-the-shelf

DS:    *Dossier de Sécurité* (Safety File)

EMC:    ElectroMagnetic Compatibility

EOQA:    *Expert ou Organisme Qualifié Agréé* (Agreed qualified expert or body)

ERTMS:[12]    European Rail Traffic Management System

EPSF:[13]    *Etablissement Public de Sécurité Ferroviaire* (Public Rail Safety Establishment)

IEC:[14]    International Electrotechnical Commission

ISA:    Independent Safety Assessor

ISO[15]:    International Organization for Standardization

E/E/PE:    *Electrique/Electronique/Programmable Electronique* (Electrics/ Electronics/ Programmable Electronics)

NoBo:    Notified Body

RAM:    Reliability Availability and Maintainability

SIL:    Safety Integrity Level

SSIL:    Software Safety Integrity Level

TSI:    Technical Specifications of Interoperability

STRMTG:[16] *Service Technique des Remontées Mécaniques et des Transports Guidés* (Technical Service for Installed Mechanics and Guided Transport)

---

11 See www.cenelec.eu.

12 See the website : http://www.ertms-online.com.

13 For more information about EPSF, see http://www.securite-ferroviaire.fr.

14 For more information, see http://www.iec.ch.

15 http://www.iso.org/iso/home.html.

16 For more information about STRMTG, see http://www.strmtg.equipement.gouv.fr.

# Conclusion

Different sectors use different standards to define the sets of techniques to be used in order to achieve and demonstrate safety. In Chapter 3, we described which standards apply to which sectors.

Let us recall that for all the domains, a number of safety analyses must be performed so as to identify the safety requirements, which are derived from the possible hazards. One of the problems at this level lies in the fact that the notions of hazard and feared event are associated with particular sectors, as each sector has specific definitions resulting from specific types of safety management. In Chapter 4, we have highlighted notions related to the rail sector.

As the equipment is made up of both hardware and software components, the starting point is a safety objective which is subsequently developed for the hardware and the software. Once the variations are obtained, it is necessary to select the various techniques to make software and hardware safe. Chapter 5 describes the techniques to achieve hardware safety, whereas Chapter 6 addresses software safety.

As far as software is concerned, it is possible to achieve safety but this will result in making the software more complex and difficult to test. This is why it is preferable, when it comes to the software, to manage the quality instead (as a minimum based on the ISO 9001:2008 standard). It is to be noted that the notion of DAL (Design Assurance Level), ASIL (Automotive SIL) and SSIL (Software Safety Integrity Level) is based on managing the quality, the competences of the people involved and the configuration (from system to software).

Certifying the product may be a requirement (mandatory certification) or an asset (voluntary certification). Chapter 7 describes the principles of certification (see EN

45011[1] and ISO 17020[2] for example) by basing itself on the railway sector. The railway sector introduces the minimum requirements, such as the notion of independent assessment, which is realized through certification so as to be able to reuse the results for several projects.

## Bibliography

[IEC 98] IEC, BS EN 45011:1998, General requirements for bodies operating product certification systems, 1998.

[ISO 08] ISO, ISO 9001:2008, Systèmes de management de la qualité - Exigence, December 2008.

[ISO 12] ISO, ISO/IEC 17020:2012, Conformity assessment – Requirements for the operation of various types of bodies performing inspection, 2012.

## Glossary

| | |
|---|---|
| ASIL: | Automotive Safety Integrity Level |
| DAL: | Design Assurance Level |
| IEC[3]: | International Electrotechnical Commission |
| ISO[4]: | International Organization for Standardization |
| SIL: | Safety Integrity Level |
| SSIL: | Software Safety Integrity Level |

---

1 The EN 45011 standard defines the general requirements relative to the bodies responsible for the certification of the products.
2 The ISO/CEI 17020 standard identifies the general criteria for the functioning of different types of inspecting bodies.
3 To learn more, please visit: http://www.iec.ch.
4 See http://www.iso.org.

# Index